Second Level
MATHS
for S1-S3

2

Author: Karen Hart
Series Consultant: Carol Lyon
Series Editor: Craig Lowther

Contents

Answers

Check your answers to this workbook online: https://collins.co.uk/pages/scottish-curriculum-free-resources-maths

1 Complete the boxes:

Write the multiple of 10 that comes either side of each number.		
6630	**6632**	6640
	6642	
	6652	

Write the multiple of 100 that comes either side of each number.		
6600	**6632**	6700
	6732	
	6832	

Write the multiple of 1000 that comes either side of each number.		
6000	**6632**	7000
	7632	

2 Rounding up or rounding down.

Draw an ⬆ or ⬇ beside each number to show if it rounds up or down.

a) 7824 to the nearest 10 ☐

b) 7824 to the nearest 100 ☐

c) 7824 to the nearest 1000 ☐

d) 5055 to the nearest 10 ☐

e) 5055 to the nearest 100 ☐

f) 5055 to the nearest 1000 ☐

3 Have these numbers been rounded to the nearest 10, 100 or 1000? Tick the correct box.

			10	100	1000
61 452	→	61 000			✔
61 452	→	61 450			
61 442	→	61 400			
61 444	→	61 440			
61 944	→	62 000			

★ **Challenge**

Think of a four-digit number that rounds up when it is rounded to the nearest 10, rounds down when it is rounded to the nearest 100 and rounds up when it is rounded to the nearest 1000. How many answers can you find?

1.2 Rounding decimal fractions to two places

1 Write or circle the number in the correct place on these number lines.

7·43

7·0   7·1   7·2   7·3   7·4   7·5   7·6   7·7   7·8   7·9   8·0

7·43
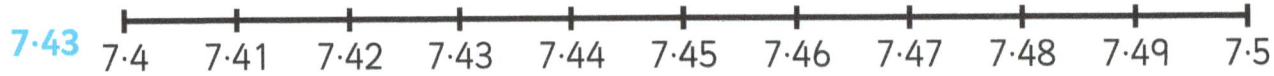
7·4   7·41   7·42   7·43   7·44   7·45   7·46   7·47   7·48   7·49   7·5

7·53
7·0   7·1   7·2   7·3   7·4   7·5   7·6   7·7   7·8   7·9   8·0

7·53
7·5   7·51   7·52   7·53   7·54   7·55   7·56   7·57   7·58   7·59   7·6

7·63

7·0   7·1   7·2   7·3   7·4   7·5   7·6   7·7   7·8   7·9   8·0

7·63
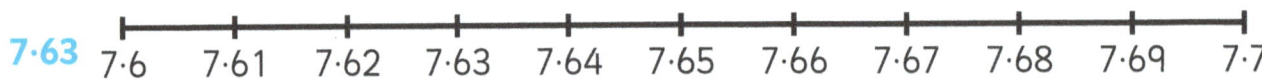
7·6   7·61   7·62   7·63   7·64   7·65   7·66   7·67   7·68   7·69   7·7

2 a) Write each decimal fraction in the correct column if they are rounded to the nearest whole number.

4·67   4·57   4·37   3·47   3·57   5·27

Rounds to 3	Rounds to 4	Rounds to 5	Rounds to 6

b) Fill in any empty boxes with numbers of your own.

3 Kaiden has been rounding decimal fractions but has made some mistakes.
Circle the answers he needs to correct.

Starting number	Nearest tenth	Nearest whole number
3·62	3·6	4
14·05	14·0	14
27·79	27·8	28
0·81	0·8	0
75·08	75·8	76
49·36	49·3	49

★ Challenge

Use the digits below to make decimal fractions with three digits that round to 5 when rounded to the nearest whole number. Each digit can be used only once.

0 1 2 3 4 5 6 7 8 9 4 5 · · · ·

Is there only one way to do this?

1.3 Using rounding to estimate the accuracy of a calculation

1 Use rounding to help you estimate an answer. Match these calculations to the most reasonable estimate.

33 + 59	100	31 + 59
37 + 59	90	35 + 59
159 – 31	130	159 – 33
159 – 35	120	159 – 37

2 Olive has £10. Can she buy chocolate for her gran, a comic for her brother, pencils for school and a bath toy for her baby cousin?

Explain your answer.

Chocolate	£ 1·49
Comic	£ 3·99
Pencils	£ 1·12
Bath toy	£ 3·55

3 Complete the table.

	Rounded to nearest thousand	Estimated answer	Rounded to nearest ten	Estimated answer
5982 + 2012	6000 + 2000	8000	5980 + 2010	7990
5982 + 3012				
5982 – 4012				

4 Use rounding to estimate each answer. Write the estimate in the correct column.

	More than 2500	Less than 2500
4631 – 1892		
4031 – 1892		
4631 – 1092		
4631 – 1992		
1992 + 1992		
1992 + 2092		
1992 + 2492		
1992 + 2692		

⭐ **Challenge**

Cerys is at the book shop. She sees two books that cost more than £3 each. She estimates that the two books will cost £7·90. What might the price of each book be? How many answers can you find?

2.1 Reading and writing whole numbers

1 Write these numbers in the place value houses using digits.

a)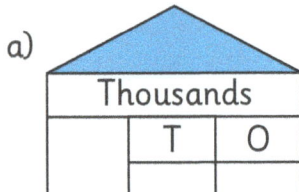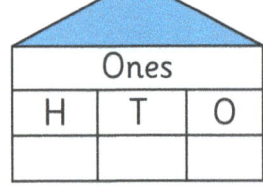

Thirty-three thousand and three

b)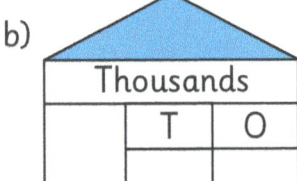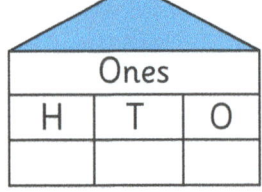

Thirty-three thousand and thirty

c)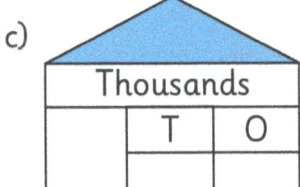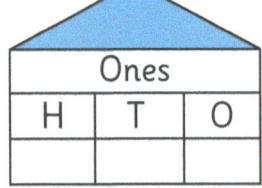

Thirty-three thousand, three hundred

d)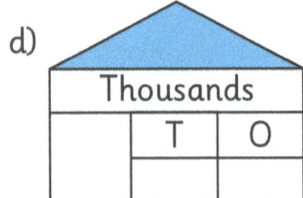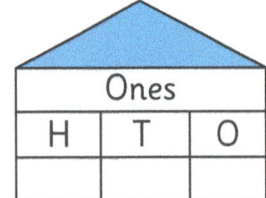

Thirty-three thousand, three hundred and three

2 Write these numbers in words.

a)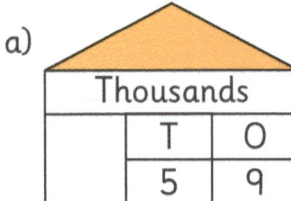

b)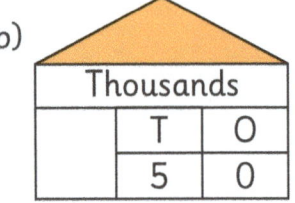

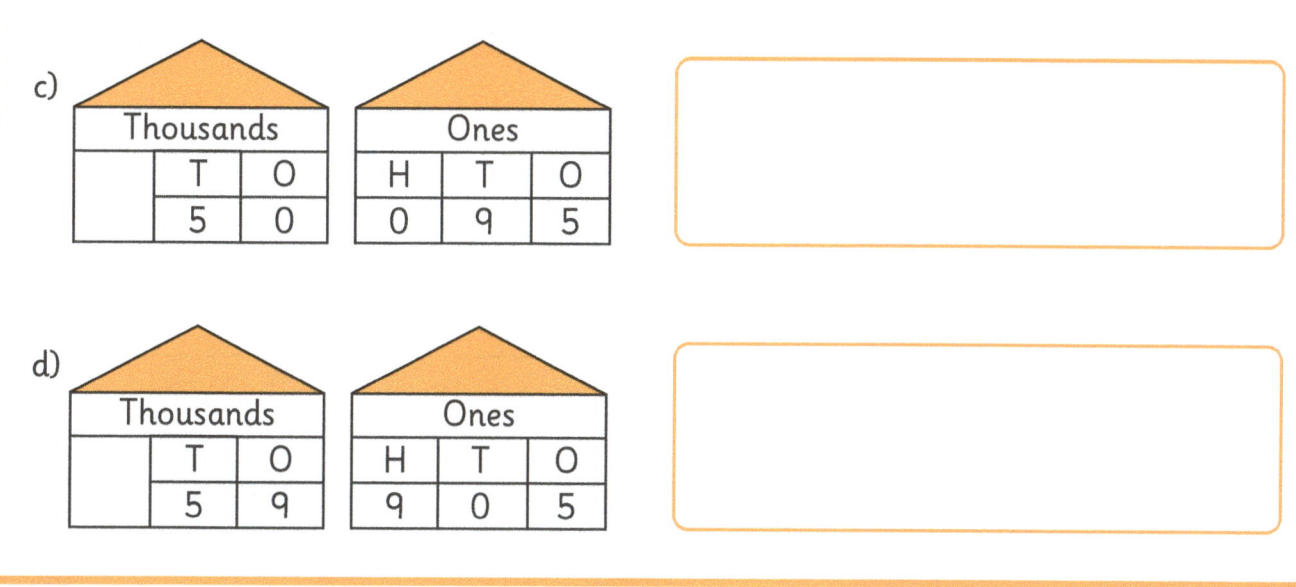

c)

Thousands		Ones		
T	O	H	T	O
5	0	0	9	5

d)

Thousands		Ones		
T	O	H	T	O
5	9	9	0	5

3 Read these numbers aloud. Mark in each time you say 'and'. The first one has been done for you.

67 5↑72   67 000   67 500   67 540   67 543   67 503

What do you notice?

Challenge

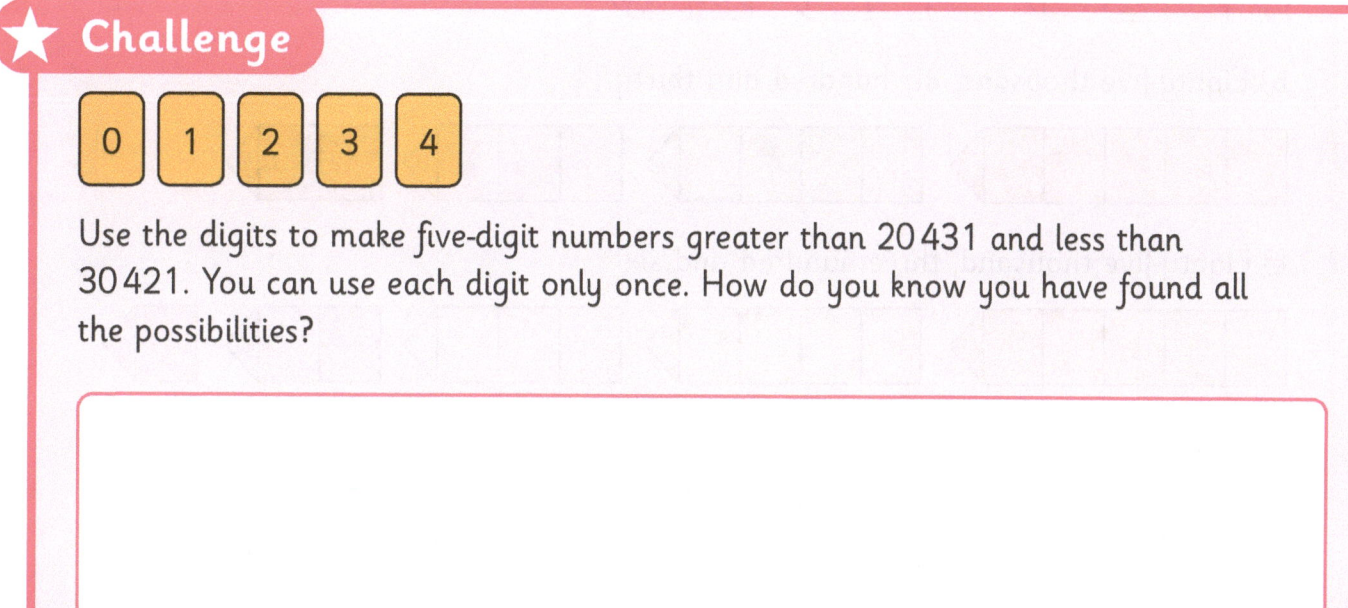

Use the digits to make five-digit numbers greater than 20 431 and less than 30 421. You can use each digit only once. How do you know you have found all the possibilities?

2.2 Representing and describing whole numbers

1 Write a five-digit number where the digit 3 has the following value:

3 ones or 3

3 tens or 30

3 hundreds or 300

3 thousands or 3000

3 tens of thousands or 30 000

2 Write the correct digits onto the place arrow cards to show each number.

a) Five thousand, six hundred and thirty

b) Eighty-five thousand, six hundred and thirty

c) Eighty-five thousand, three hundred and six

d) Eighty-five thousand and thirty-six

e) Eighty thousand and sixty

3 Complete the Think Board for the number 61 043.

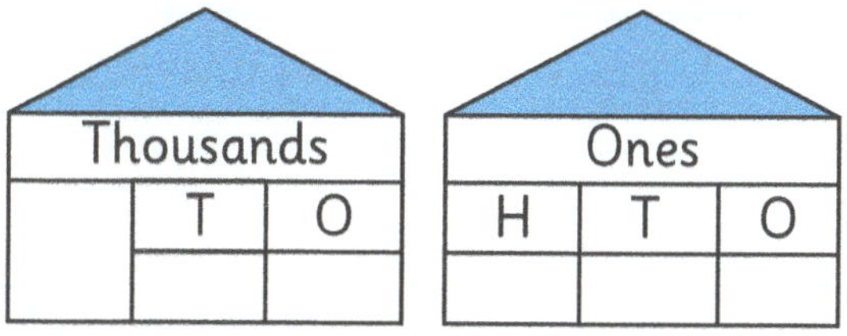

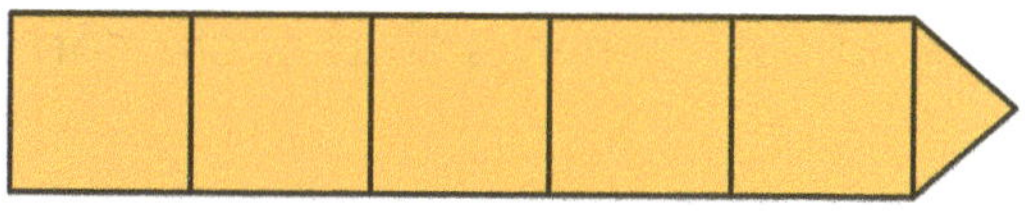

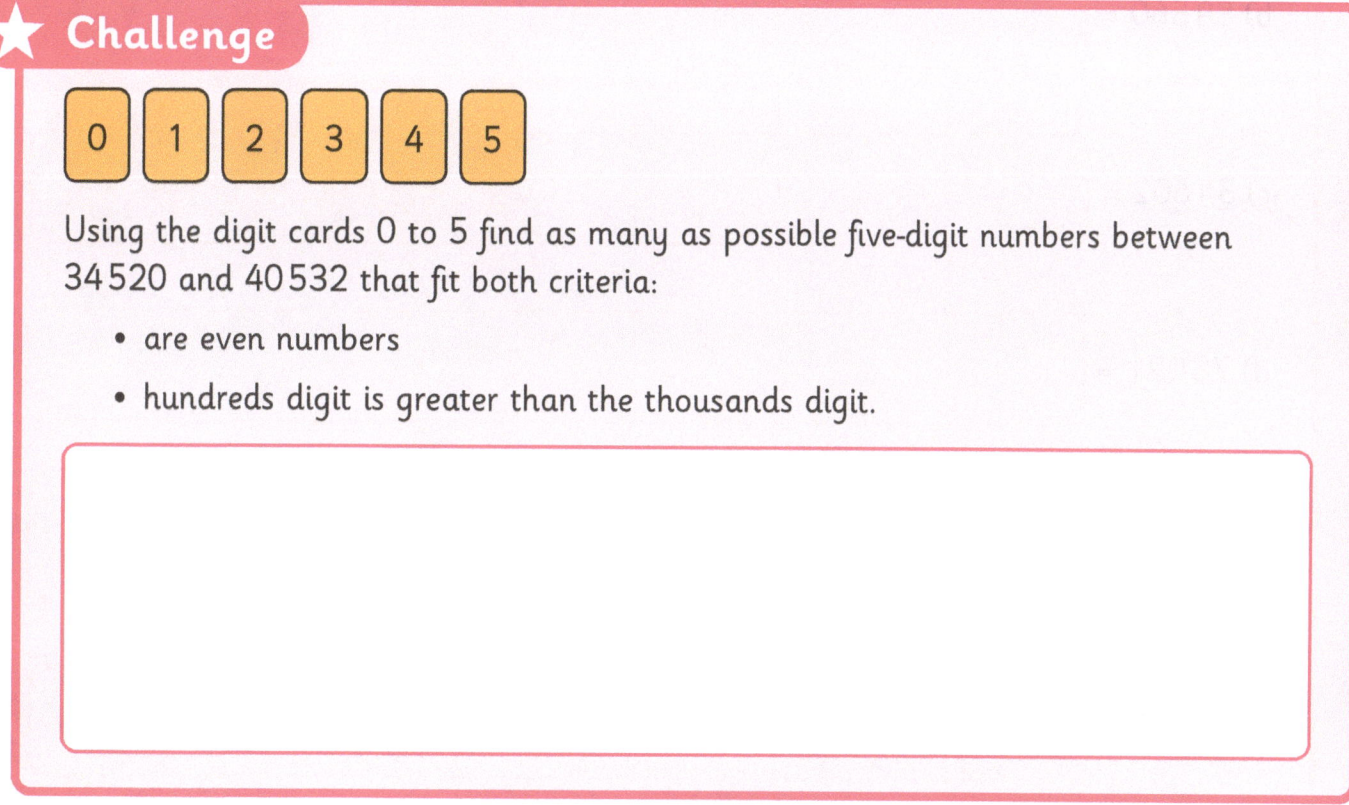

Thousands		Ones		
T	O	H	T	O

61 043

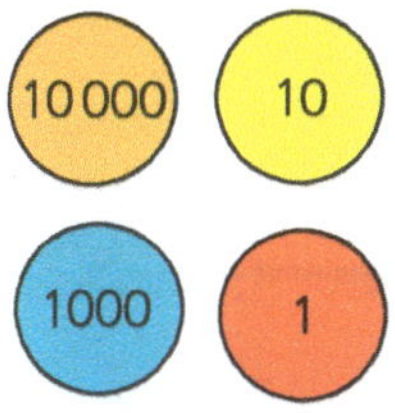

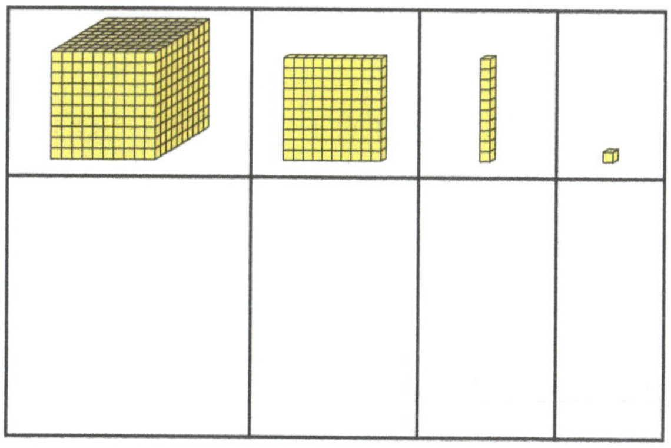

⭐ **Challenge**

0	1	2	3	4	5

Using the digit cards 0 to 5 find as many as possible five-digit numbers between 34 520 and 40 532 that fit both criteria:

- are even numbers
- hundreds digit is greater than the thousands digit.

2.3 Place value partitioning of whole numbers

1 Tick the statements that show correct partitions. Correct any wrong answers.

a) 5670 = 5000 + 600 + 70

b) 7979 = 7000 + 97 + 9

c) 34 612 = 30 000 + 4000 + 600 + 2

d) 62 523 = 60 000 + 2000 + 500 + 20 + 3

2 Partition these numbers One has been done for you.

a) 23 875 = 20 000 + 3000 + 800 + 70 + 5

b) 89 560 =

c) 54 602 =

d) 78 021 =

e) 60 532 =

f) 60 050 =

3 Complete the table. What is the bold digit worth? One is done for you.

6**6**666		6000			
66666					
6666**6**					
666**6**6					
66**6**66					

4 What number is shown here? Fill in the first column. One is done for you.

13579	10 000	3000	500	70	9
	10 000	4000	500	70	9
	10 000	4000	400	70	9
	10 000	4000	400	80	9
	10 000	4000	400		9

5 Complete the table. Can you find four different ways to make these numbers? Fill in the blanks.

	Thousands	Hundreds	tens	ones
1465		14		
				65
	1			
			146	
2576		25		
				76
	2			
			257	
3687		36		
				87
	3			
			368	

You will need a dice.

1. Roll the dice and choose one square to fill in. Keep rolling the dice and filling in a square every time.

   • Try to make Rosa have the greatest number.
   • Try to make Jay have the lowest number.
   • Try to make Amman be closest to 25 000.
   • Try to make Layla be closest to 30 000.

Jay

Rosa

Amman

Layla

2. Write about how well your final numbers fit the rules.

2.4 Number sequences

1 Keep the sequences going by counting in 10s.

a) 30, 40, _____, _____, _____, _____, _____, _____

b) 37, 47, _____, _____, _____, _____, _____, _____

c) 237, 247, _____, _____, _____, _____, _____, _____

d) 307, 297, _____, _____, _____, _____, _____, _____

e) 1307, 1297, _____, _____, _____, _____, _____, _____

2 Complete the blanks.

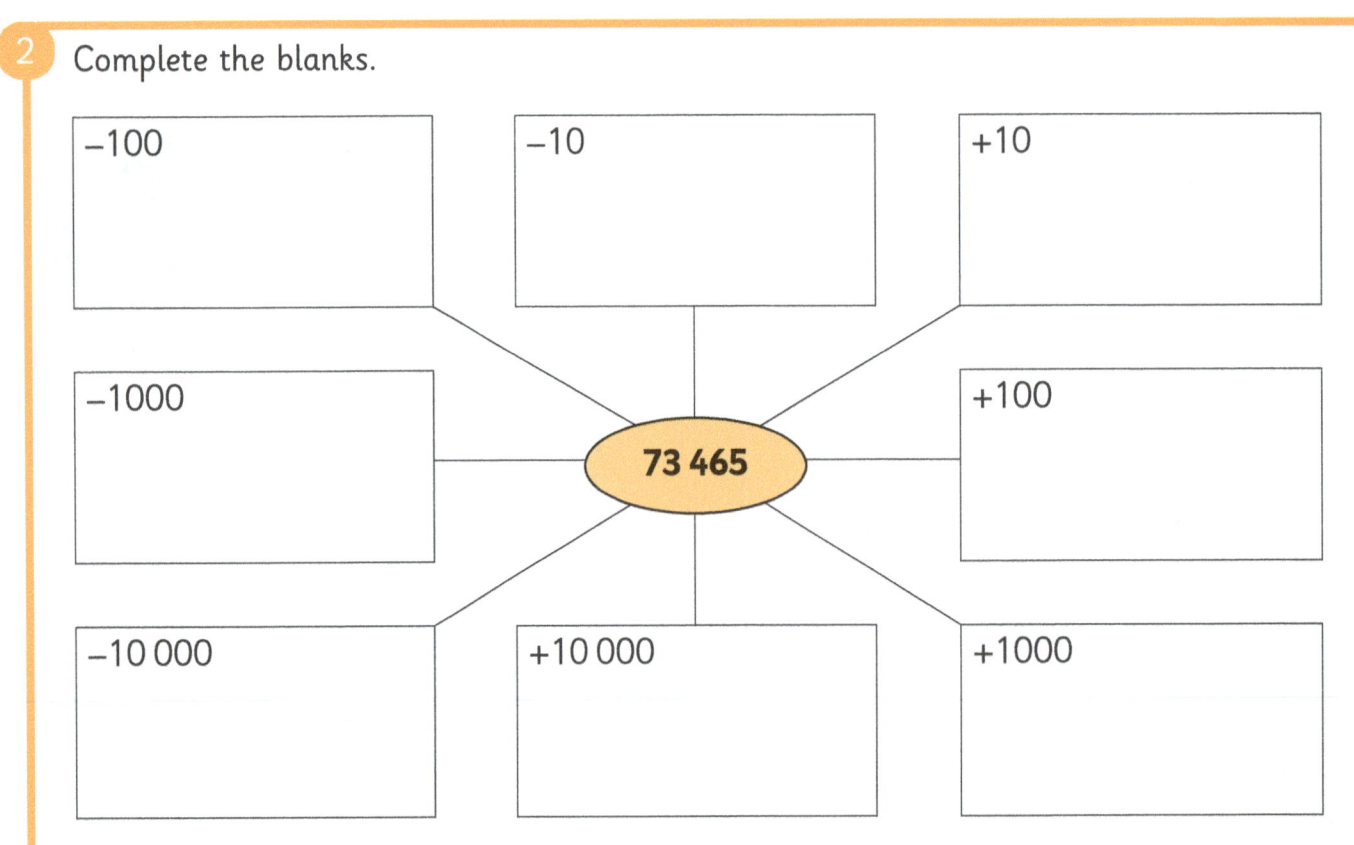

3 Work out these number sentences:

a) 38 890 + 10 = ☐

b) 38 990 + 10 = ☐

c) 38 890 + 100 = ☐

d) 38 990 + 100 = ☐

e) 38 890 + 1000 = ☐

f) 39 990 + 1000 = ☐

4 What has been added or subtracted to these numbers to make them balance? One has been done for you.

a) 24 506 ➡ 24 406 [− 100]

b) 34 506 ➡ 34 606 ☐

c) 34 506 ➡ 34 406 ☐

d) 34 506 ➡ 33 506 ☐

e) 34 506 ➡ 34 496 ☐

f) 34 156 ➡ 34 056 ☐

⚀	add 10
⚁	subtract 10
⚂	add 100
⚃	subtract 100
⚄	add 1000
⚅	subtract 1000

1. Choose a five-digit number between 2000 and 8000 for Anu and a different number for Archie. Write it in the table.

2. Roll a dice and use the code. Start with Anu and then Archie. When you have finished, decide who is closest to their start number.

3. Choose new start numbers and repeat the game. How many times does Anu win?

Anu						
Archie						

Anu						
Archie						

Anu						
Archie						

Anu						
Archie						

Anu						
Archie						

1 Write the number that the first digit is worth in each of these. One has been done for you.

a) 65 000 ➤ 60 000

b) 650 ➤ _____

c) 6500 ➤ _____

d) 5600 ➤ _____

e) 56 000 ➤ _____

Now put the answers in order from smallest to largest.

_____ _____ _____

_____ _____

2 Complete these statements using < or >.

a) 21 087 ☐ 21 097

b) 21 087 ☐ 21 084

c) 21 187 ☐ 20 087

d) 21 087 ☐ 20 987

e) 21 087 ☐ 21 807

f) 20 087 ☐ 20 870

3   The table shows spectators at football matches on one Sunday in May.

Team	Spectators
Glenfield United	4845
Hillside Albion	40845
Riverdale F.C.	4895
Shore United	1845
Rockvalley F.C.	14805
Woodburn City	20895

a) Write the names of the teams in order from largest numbers of spectators to smallest.

b) What is the difference between the smallest and largest crowd numbers?

4 Complete these to make them true.

a) 34 506 > | 3 | 4 | | | 6 |

b) 34 506 > | 3 | | | 0 | 6 |

c) 34 506 < | | | 5 | 0 | 6 |

d) | 3 | | | | 6 | < 30 606

e) | 3 | | | | 6 | > 39 006

★ **Challenge**

Use the digits 0 – 9 to make each statement true. You can only use each digit once. How many different ways can you do it?

| 0 | 1 | 2 | 3 | 4 | 5 | 6 | 7 | 8 | 9 |

| | | | | | < 81 605

| | | | | | > 19 495

1

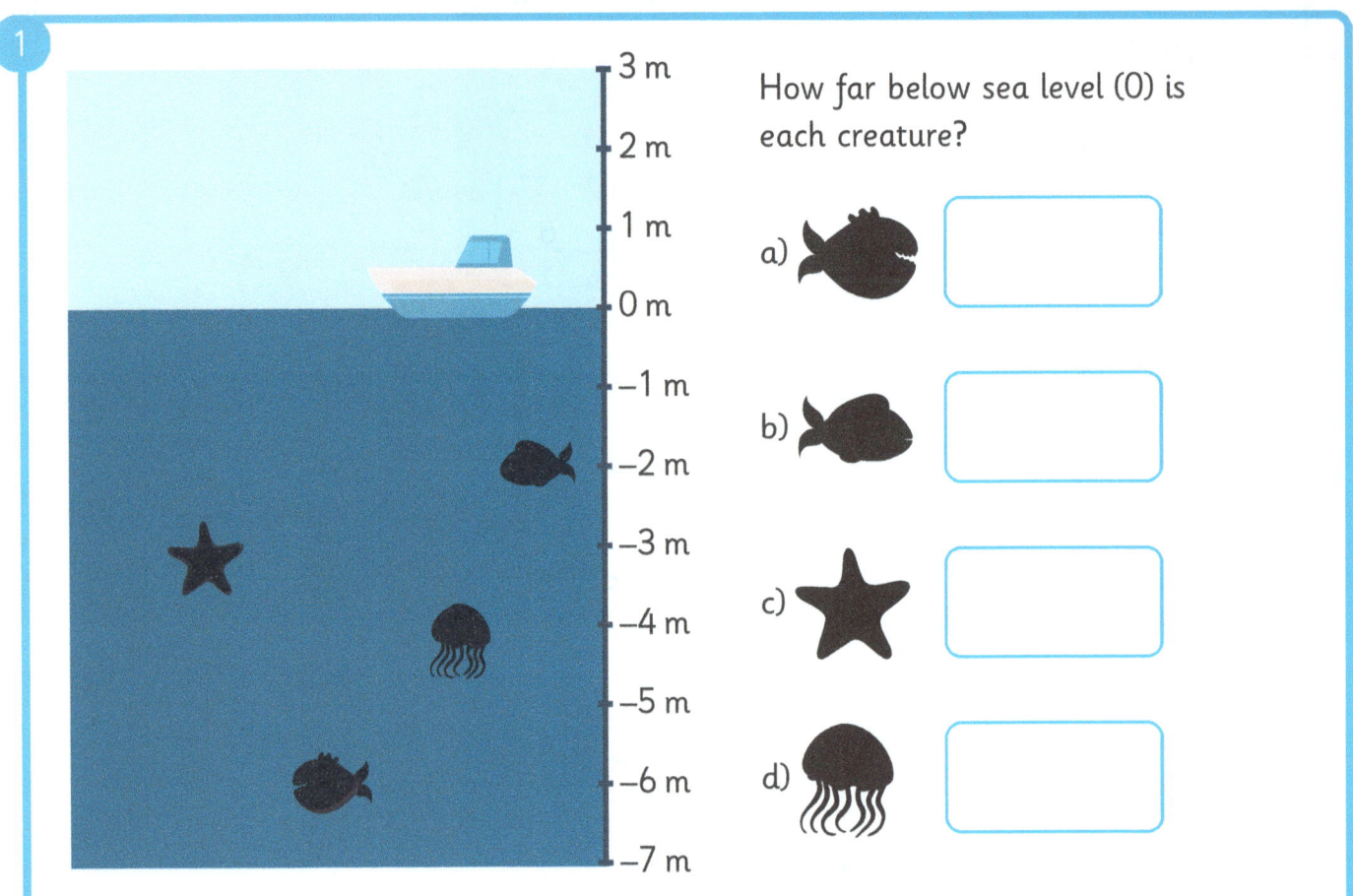

How far below sea level (0) is each creature?

a) [ ]

b) [ ]

c) [ ]

d) [ ]

2 Look at these temperatures. Circle the temperature which is colder.

a) 10°C or 12°C

b) 0°C or 2°C

c) –2°C or 0°C

d) –12°C or 10°C

e) –49°C or –19°C

f) 49°C or 19°C

g) –49°C or 19°C

h) 49°C or –19°C

i) –4°C or –40°C

3 Keep these sequences going.

a) 4, 3, 2, 1, 0, –1, _____, _____, _____, _____

b) 40, 30, 20 _____, _____, _____, _____, _____

c) –7, –6, _____, _____, _____, _____, _____, _____

d) –70, –60, _____, _____, _____, _____, _____

4 Use < or > to make these statements true.

a) –6 ⬚ –7

b) –6 ⬚ 7

c) 6 ⬚ –7

d) 0 ⬚ –7

e) –6 ⬚ 0

f) –60 ⬚ –70

⭐ **Challenge**

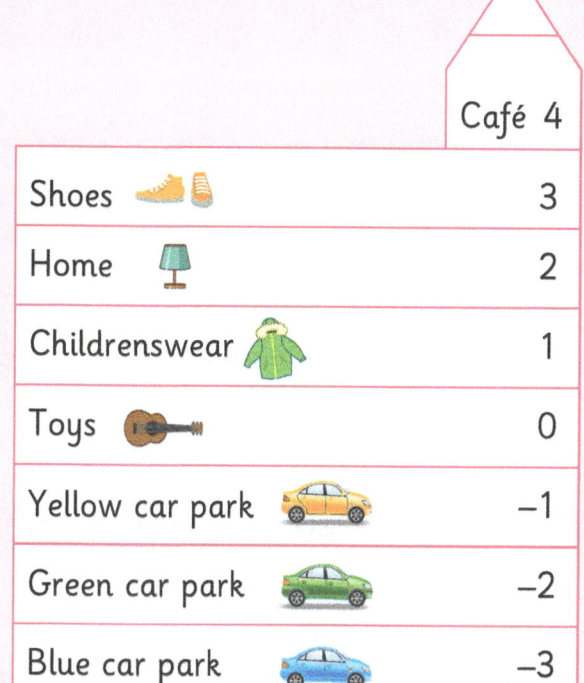

Café	4
Shoes	3
Home	2
Childrenswear	1
Toys	0
Yellow car park	–1
Green car park	–2
Blue car park	–3

James takes the lift in the shopping centre and goes up three floors. Write four possible lift journeys where he might have started and then got out of the lift.

2.7 Reading and writing decimal fractions

1   a) What decimal fraction does each diagram show?

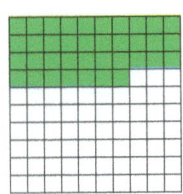   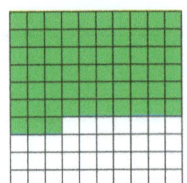

b) Colour the diagrams to show the following decimal fractions.

i)    Show 0·31

ii) 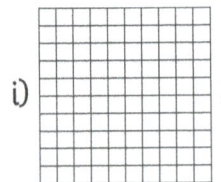   Show 1·31

iii)    Show 1·03

2   Circle the odd one out:

a) 9·06   nine point six $9\frac{6}{100}$

b) 9·6   nine point six $9\frac{6}{100}$

c) 9·61   nine point one $9\frac{61}{100}$

d) 6·09   six point nine $6\frac{9}{100}$

e) 6·99   six point ninety-nine $6\frac{9}{100}$

f) 6·9   six point nine $6\frac{9}{100}$

Complete the table.

Words	Diagram	Decimal fraction	Mixed number
One point seven five			
			$2\frac{55}{100}$

★ Challenge

What number does each student have?

Thomas: My number has 6 tenths and 5 hundredths.

My number has 3 more tenths and 4 more hundredths than Thomas.

My number has 5 more hundredths than Thomas.

My number has 7 more tenths than Thomas.

2.8 Representing and describing decimal fractions

1 Colour the grids four different ways to show 0·45.

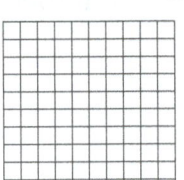 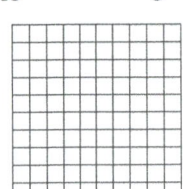 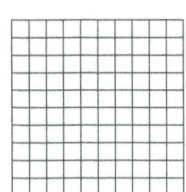 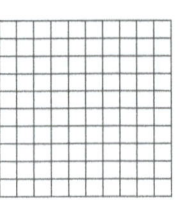

2 Complete the table. The first one is done for you.

Diagram	Decimal fraction (shaded part)	Decimal fraction (unshaded part)	Fraction (shaded part)
	0·25	0·75	$\frac{25}{100}$

3 Match the decimal fraction to the fraction.

a) 1·34

b) 1·04

c) 1·3

d) 34·1

e) 34·01

$1\frac{30}{100}$

$1\frac{4}{100}$

$1\frac{34}{100}$

$34\frac{10}{100}$

$34\frac{1}{100}$

★ **Challenge**

How many different numbers can you make using these place value arrows? You can use 1, 2 or 3 arrows each time.

6 0·3 0·08

2 0·7 0·04

29

1 Write these decimal fractions in the grid below. Think about where each number can go.
Write them from largest to smallest.

9·06   9·86   36·69   13·8   9·65   9·6   24·17

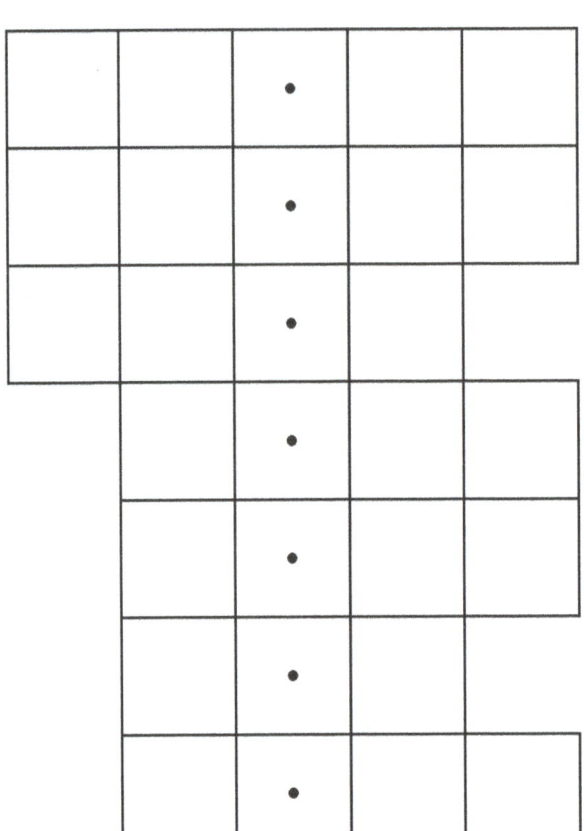

2 Use the symbols =, < or > to compare these decimal fractions.

a) 27·17 ☐ 27·19

b) 27·07 ☐ 20·77

c) 15·45 ☐ 15·55

d) 15·5 ☐ 15·05

e) 60·9 ☐ 6·95

f) 60·90 ☐ 60·9

3 Write the numbers from largest to smallest.

a) 0·1   0·15   0·01   _____   _____   _____

b) 1·67   1·7   1·77   _____   _____   _____

c) 28·04 28   28·4   _____   _____   _____

d) 0·50   5·0   0·05   _____   _____   _____

e) 4   4·1   4·01   _____   _____   _____

★ Challenge

1. For each decimal fraction, roll a dice and write a digit into one of the empty boxes. Keep rolling and filling in the empty boxes. Can you make six decimal numbers that are all less than 60?

a) [ ][ ] · [ ][ ]   b) [ ][ ] · [ ][ ]

c) [ ][ ] · [ ][ ]   d) [ ][ ] · [ ][ ]

e) [ ][ ] · [ ][ ]   f) [ ][ ] · [ ][ ]

2. List your numbers from smallest to largest.

3. Try again but this time try to make all the numbers less than 30.

a) [ ][ ] · [ ][ ]   b) [ ][ ] · [ ][ ]

c) [ ][ ] · [ ][ ]   d) [ ][ ] · [ ][ ]

e) [ ][ ] · [ ][ ]   f) [ ][ ] · [ ][ ]

3.1 Mental addition and subtraction

1 Calculate the following:

a) 156 – 30 = [ ]

b) 136 + 30 = [ ]

c) 687 – 120 = [ ]

d) 567 + 120 = [ ]

e) 503 – 40 = [ ]

f) 503 + 40 = [ ]

g) 758 + 70 = [ ]

h) 758 – 70 = [ ]

2 Transform these calculations. Two have been done for you.

a) 134 + 159 = [133 + 160]

b) 144 + 149 = [ ]

c) 698 + 36 = [ ]

d) 697 + 136 = [ ]

e) 159 – 42 = [160 – 43]

f) 158 – 62 = [ ]

g) 477 – 139 = [ ]

h) 809 – 49 = [ ]

3 Transform these numbers and calculate mentally. Jot down numbers that help. The first one is done for you.

| 607 + 269
606 + 270
806 + 70
876 | 607 + 267 | 499 – 46 | 499 – 48 |

| 698 + 146 | 696 + 148 | 701 – 54 | 703 – 54 |

⭐ **Challenge**

You are only allowed to use the digits 8, 9 and 0 for this challenge. Create a four-digit plus three-digit sum. Transform and solve the question.

☐ ☐ 9 ☐ + ☐ ☐ ☐

= ☐ ☐ ☐ 0 + ☐ ☐ ☐

= ☐ ☐ ☐ ☐

3.2 Adding and subtracting a string of numbers

1 Circle the numbers in each question that total a multiple of 100. One has been done for you.

Space for jottings

a) (4550) + 3423 + (150)

b) 3460 + 1240 + 3016

c) 230 + 3044 + 3070

d) 1580 + 1520 + 1582

e) 710 + 5602 + 490

2 Circle the numbers which total a multiple of 10 or 100 and calculate mentally. Jot down numbers if it helps. The first one is done for you.

Space for jottings

a) 6523 − (179) − (221) = | 6123 |

b) 6523 − 189 − 211 =

c) 6803 − 189 − 211 =

d) 6803 − 355 − 245 =

e) 6803 − 155 − 145 =

3 Look for multiples of 10, 100 or 1000 and calculate mentally. One has been done for you.

(670) + 1026 + (230) = 1026 + 900 1926	5740 – 290 – 410 =
6522 – 2500 – 1500 =	7604 + 196 + 177 =
1045 – 307 – 23 =	644 + 466 + 156 =

★ Challenge

1. Write pairs of numbers that help you add or subtract these calculations mentally. Can you find different pairs that help? The first one has been done for you.

   a) **934 + 766 + 104**   766 + 104 = 870   34 + 66 = 100

   b) 547 + 1453 + 107

   c) 2006 + 1994 + 594

2. Now create two additions of your own with numbers that total multiples of 10 or 100 in them.

3.3 Adding and subtracting multiples of 10, 100 and 1000

1 Partition each number into its place values. The first one is done for you.

a) 14567 ➡ 14000 + 500 + 60 + 7

b) 14507 ➡

c) 15067 ➡

d) 15007 ➡

e) 15670 ➡

2 Use partitioning to work these out. Look at the example first.

21342 + 4526 ➡ 21000 + 300 + 40 + 2 + 4000 + 500 + 20 + 6 ➡
25000 + 800 + 60 + 8 = 25868

a) 10546 + 2522

b) 10546 + 354

c) 454 + 20 546

d) 20 546 − 445

e) 20 566 − 556

3 Use place value to calculate. One has been done for you.

a) 36 278 + 4321

Thousands		Ones		
T	O	H	T	O
3	6	2	7	8

+

Thousands		Ones		
T	O	H	T	O
	4	3	2	1

Thousands		Ones		
T	O	H	T	O
4	0	5	9	9

36 thousands + 4 thousands equals 40 thousands

2 hundreds + 3 hundreds equals 5 hundreds

7 tens add 2 tens equals 9 tens

8 ones and 1 one equals 9 ones

b) 40 599 – 4321

Thousands			Ones		
T	O		H	T	O
4	0		5	9	9

–

Thousands			Ones		
T	O		H	T	O
	4		3	2	1

Thousands			Ones		
T	O		H	T	O

c) 40 599 – 5421

Thousands			Ones		
T	O		H	T	O
4	0		5	9	9

–

Thousands			Ones		
T	O		H	T	O
	5		4	2	1

Thousands			Ones		
T	O		H	T	O

d) 5491 + 23218

Thousands	
T	O
	5

Ones		
H	T	O
4	9	1

+

Thousands	
T	O
2	3

Ones		
H	T	O
2	1	8

Thousands	
T	O

Ones		
H	T	O

★ **Challenge**

Fill in the blanks to make each statement true.

1 ☐ 3 2 ☐ + ☐ ☐ 5 5 = 2 4 6 ☐ 6

8 0 ☐ 9 8 – 1 ☐ 4 ☐ 4 = 6 ☐ 3 5 4

1 Add these numbers by partitioning. The first one is done for you. Remember to line up your columns.

a)
```
    3 4 4 2 3
  + 1 2 5 7 2
  ───────────
    4 0 0 0 0
      6 0 0 0
        9 0 0
          9 0
            5
  ───────────
    4 7 9 9 5
```

b)
```
    3 4 4 2 3
  + 1 2 5 2 7
  ───────────
```

c)
```
    3 4 4 3 3
  + 1 6 5 2 7
  ───────────
```

d)
```
    1 6 5 2 7
  + 3 4 5 2 7
  ───────────
```

2 Subtract these numbers by partitioning. The first one is done for you.

a)

	7	8	6	3	7
−	6	5	5	1	4
	1	0	0	0	0
		3	0	0	0
			1	0	0
				2	0
					3
	1	3	1	2	3

b)

	7	8	6	3	7
−	5	5	5	2	4

c)

	7	8	6	3	7
−	3	5	5	3	4

d)

	7	8	6	3	7
−	3	5	6	2	7

3 Use partitioning to calculate the following:

a) 8 6 6 5 7 − 7 5 4 4 4

b) 7 6 5 4 6 − 6 5 4 4 4

c) 5 1 0 2 + 6 4 9 1 8

d) 6 4 9 0 8 + 5 1 0 2

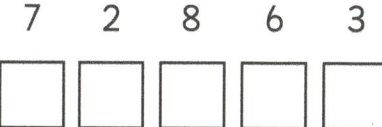

4 Fill in the missing numbers to make each calculation correct.

a)
```
     7   2   8   6   3
  -  □   □   □   □   □
  ─────────────────────
     5   1   1   3   3
  ─────────────────────
```

b)
```
     7   2   8   6   3
  +      □   □   □   □
  ─────────────────────
     7   8   9   9   7
  ─────────────────────
```

⭐ **Challenge**

Choose five of the following numbers. Write them out like the partitions in question 1 and work out what your starting numbers are.

6000

20 000

10

70

1000

7

500

2

+

43

1 Complete the following number sentences. The first has been done for you.

a) 26 + 5 = 2 tens + 6 ones + 5 ones. This is 2 tens + 1 ten + 1 one.

b) 46 + 7 = 4 tens + 6 ones + 7 ones.

This is

c) 460 + 70 = 4 hundreds + 6 tens + 7 tens.

This is

d) 460 + 74 = 4 hundreds + 6 tens + 7 tens + 4 ones.

This is

e) 740 + 46 = 7 hundreds + 4 tens + 4 tens + 6 ones.

This is

f) 46 + 74 = 4 tens + 6 ones + 7 tens + 4 ones.

This is

g) 460 + 730 = 4 hundreds + 6 tens + 7 hundreds + 3 tens.

This is

2 Use column addition to work these out.

a)
```
   4 3 5 6
 + 2 3 2 7
 _____
```

b)
```
   4 3 5 6
 + 2 3 7 2
 _____
```

c)
```
   4 3 5 6
 + 2 7 3 2
 _____
```

d)
```
   4 3 5 6
 + 7 2 3 2
 _____
```

e)
```
   1 0 0 6
   4 5 0 6
 +    5 0
 _____
```

f)
```
   2 0 0 6
   3 5 0 6
 +   5 0 0
 _____
```

g)
```
   2 5 0 6
   3 5 0 6
 +       5
 _____
```

h)
```
   2 5 6 0
   2 0 6 0
 + 5 0 0 0
 _____
```

3 Use column addition to work these out.

a) 6207 + 544 + 291   b) 654 + 201 + 6217   c) 7328 + 312 + 65

⭐ **Challenge**

The answer is 1368.

| 1 | 2 | 3 | 4 | 5 | 6 | 7 | 8 | 9 |

1. Use the digits 1 to 9 to create an addition calculation that gives this answer. You can only use each digit once.

+

    1   3   6   8

2. Use the digits 1–9 to create a different addition calculation that gives this answer. You can only use each digit once.

+

    1   3   6   8

3. Can you find more possible answers?

45

1 Complete the calculations. The first one is done for you. You could use a different coloured pencil to show your negative numbers.

a)
```
    7 1 6
  - 3 4 2
  -------
    4 0 0
  -  3 0
  -------
        4
  -------
    3 7 4
```

b)
```
    7 7 1
  - 2 3 4
  -------


  -------

```

c)
```
    5 0 8
  - 1 7 4
  -------


  -------

```

d)
```
    9 4 0
  - 5 3 7
  -------


  -------

```

2 Check the answers to these subtractions. Tick them if they are correct.

a)
```
    8 0 8
  - 5 9 5
  -------
    3 0 0
  -  1 0
  -------
        3
  -------
    3 1 3
```

b)
```
    7 0 7
  - 4 9 5
  -------
    3 0 0
  -  9 0
  -------
        2
  -------
    2 1 2
```

c)
```
    6 8 0
  - 1 0 4
  -------
    5 0 0
  -  8 0
  -------
        6
  -------
    4 2 6
```

d)
```
    5 8 0
  - 2 2 9
  -------
    3 0 0
       6 0
  -    9
  -------
    2 3 9
```

e)
```
    5 8 0
  - 1 9 3
  -------
    4 0 0
  -  1 0
  -    3
  -------
    3 8 7
```

3 Use the space below to correct the wrong answers in question 2.

828

174

368

753

Choose two numbers from the list above to create a subtraction calculation that gives:

1. The largest possible answer.

2. The smallest possible answer.

3. The answer that is a multiple of 10.

1 Match each calculation to its answer.

a) 600 − 400

b) 900 − 500

200

c) 700 − 500

300

d) 800 − 400

400

e) 600 − 300

f) 800 − 500

2 Complete the calculations. The first one is done for you. You could use a different coloured pencil to show your negative numbers.

a)
```
    3 6 9 2
  − 1 8 8 0
  ─────────
    2 0 0 0
    − 2 0 0
        1 0
         2
  ─────────
    1 8 1 2
```

b)
```
    3 6 8 2
  − 1 8 8 0
  ─────────

  ─────────
```

c)
```
    3 6 8 2
  − 1 5 8 8
  ─────────

  ─────────
```

d)
```
    8 6 8 2
  − 1 5 9 1
  ─────────

  ─────────
```

e)
```
    8 6 8 2
  − 1 5 9 5
  ─────────

  ─────────
```

3 Complete the missing answers in these number spiders.

a)

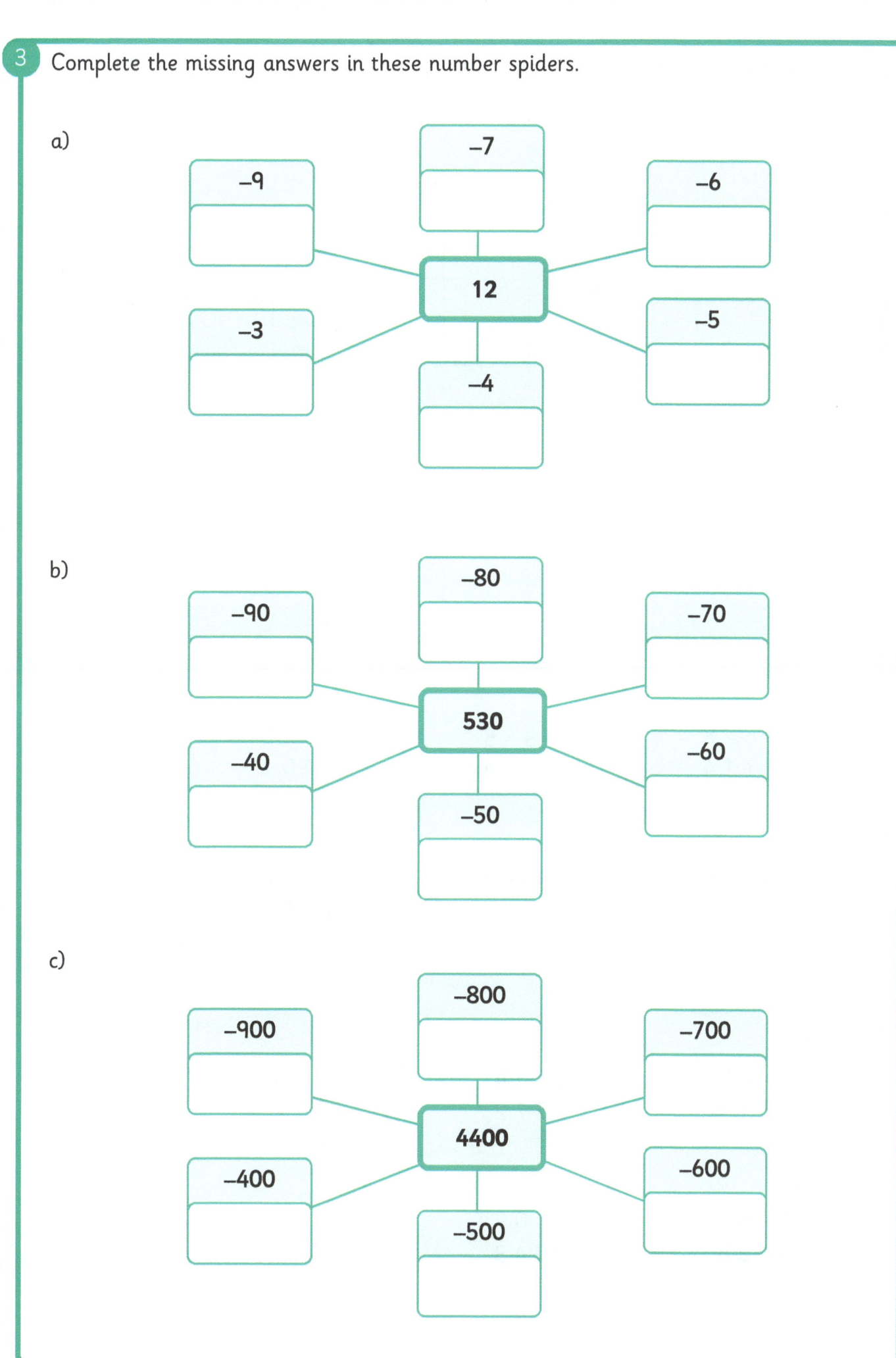

4 Complete the calculations and complete the puzzle.

1	▨	2	3	4	
	▨	5			▨
6	7			▨	8
9		▨	▨	10	
▨	11		▨		
12			▨		

Across

2.   3628
  − 1714

5.   4365
  − 3644

6.   4529
  − 2901

9.   5637
  − 5595

10.   1360
  − 1279

11.   7064
  − 6173

12.   5066
  − 1039

Down

1.   8228
  − 1714

2.   6944
  − 6772

3.   9039
  − 8111

4.   8002
  − 7991

7.   8968
  − 2688

8.   6509
  − 2343

Create 4 subtraction calculations so that the puzzle is correct.

¹4			
²3	2	³6	
8		3	
⁴2	7	8	2

Across

2.

4.

Down

1.

3.

3.8 Subtracting three-digit numbers using standard algorithms

1 Use a standard algorithm to work these out.

a)
```
   3 6 6
 - 1 2 7
 _____
```

b)
```
   3 6 6
 - 1 2 9
 _____
```

c)
```
   3 9 6
 - 1 2 8
 _____
```

d)
```
   3 9 6
 - 1 4 7
 _____
```

e)
```
   5 7 4
 - 2 4 5
 _____
```

f)
```
   5 7 6
 - 2 4 8
 _____
```

g)
```
   5 7 8
 - 3 4 9
 _____
```

h)
```
   5 8 0
 - 3 4 9
 _____
```

2 Use a standard algorithm to work these out.

a)
```
   8 2 7
 - 5 5 5
 _____
```

b)
```
   8 2 7
 - 4 4 4
 _____
```

c)
```
   8 2 7
 - 3 3 3
 _____
```

d)
```
   8 0 3
 - 3 3 3
 _____
```

e)
```
   9 4 9
 - 4 9 7
 _____
```

f)
```
   9 4 9
 - 4 8 7
 _____
```

g)
```
   9 4 9
 - 4 5 8
 _____
```

h)
```
   9 3 9
 - 4 5 8
 _____
```

3 Use a standard algorithm to work these out.

a)
```
   7 1 3
 - 5 2 5
 _____
```

b)
```
   7 1 3
 - 5 3 5
 _____
```

c)
```
   7 1 3
 - 5 4 6
 _____
```

d)
```
   7 0 3
 - 5 4 6
 _____
```

e)
```
   6 6 3
 - 2 7 7
 _____
```

f)
```
   6 5 3
 - 2 7 7
 _____
```

g)
```
   6 4 3
 - 2 8 7
 _____
```

h)
```
   6 3 3
 - 2 9 7
 _____
```

| 0 | 1 | 2 | 3 | 4 | 5 |

1) Use the digits 0 to 5 to create calculations where you have to exchange twice.
   Find the answers to your calculations.

| ☐ ☐ ☐ ☐ | ☐ ☐ ☐ | ☐ ☐ ☐ ☐ |
| – ☐ ☐ ☐ | – ☐ ☐ ☐ | – ☐ ☐ ☐ |

2) Create three subtraction questions where the answer is always 5555.
   The first question should have no exchange, the second 1 exchange and the third 2 exchanges.

| ☐ ☐ ☐ ☐ | ☐ ☐ ☐ ☐ | ☐ ☐ ☐ ☐ |
| – ☐ ☐ ☐ ☐ | – ☐ ☐ ☐ | – ☐ ☐ ☐ |

1 Write the number shown here.

a) | 3 hundreds | 40 ones |

b) | 3 hundreds | 4 tens |

c) | 3 hundreds | 40 tens |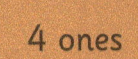

d) | 3 hundreds | 4 tens | 4 ones |

e) | 3 hundreds | 4 tens | 40 ones |

f) | 3 hundreds | 40 tens | 40 ones |

2 Find the incorrect calculations. Draw a ring around them.

a)
$$\begin{array}{r} {}^{3}4\,{}^{1}7\,{}^{2}\!8\,{}^{1}4 \\ -\ 1\,9\,2\,7 \\ \hline 2\,8\,0\,7 \end{array}$$

b)
$$\begin{array}{r} {}^{3}4\,{}^{1}8\,{}^{2}\!8\,{}^{1}4 \\ -\ 2\,9\,2\,7 \\ \hline 1\,7\,0\,7 \end{array}$$

c)
$$\begin{array}{r} 4\,6\,{}^{1}\!8\,{}^{1}3 \\ -\ 1\,7\,1\,4 \\ \hline 3\,1\,0\,9 \end{array}$$

d)
$$\begin{array}{r} {}^{3}\!4\,{}^{1}0\,2\,3 \\ -\ 1\,7\,1\,2 \\ \hline 2\,3\,1\,1 \end{array}$$

3 Calculate these:

a) 5 3 1 6
   – 2 4 7 2
   _____

b) 5 3 0 6
   – 2 5 7 2
   _____

c) 6 3 0 5
   – 3 5 8 8
   _____

d) 6 3 1 5
   – 3 5 8 8
   _____

e) 8 0 1 2
   – 3 4 1 1
   _____

f) 8 1 0 2
   – 3 4 1 1
   _____

g) 8 1 2 0
   – 3 4 1 1
   _____

h) 8 0 0 2
   – 3 4 1 1
   _____

⭐ **Challenge**

Fill in the missing digits to make each subtraction correct.

1.   ☐ 8 ☐ 7
   –   4 ☐ 1 ☐
   _____
     1 9 1 9

2.   5 7 3 7
   – ☐ ☐ ☐ ☐
   _____
     1 9 1 9

3.10 Mental and written strategies

1 Using these methods, solve this calculation **1569 + 546**.

a) An empty number line.

├───┤

1569

b) Partitioning.

_____ + _____ + _____ + _____ + _____ + _____

c) Column partitioning.

```
   1 5 6 9
+    5 4 6
```

d) Standard algorithm

```
   1 5 6 9
+    5 4 6
```

2 Use the cards to create three addition calculations that are most efficiently solved as mental calculations.

3600   3876   5397

2040   2667   1199

_____ + _____ = _____

_____ + _____ = _____

_____ + _____ = _____

	Distance in kilometres
Inverness to Aberdeen	166
Edinburgh to Glasgow	75
Glasgow to Inverness	270
Dundee to Edinburgh	100
Aberdeen to Dundee	106

How far is it to get from Inverness to Glasgow, passing through Aberdeen, Dundee and Edinburgh?

Write down your method.

⭐ **Challenge**

Cameron adds two four-digit numbers to make a total of 11 074. He only uses the digits 3 and 7.

a) What are his two numbers?

```
  ☐ ☐ ☐ ☐
+ ☐ ☐ ☐ ☐
_____

_____
```

b) If Cameron created a subtraction using his two numbers, what would the answer be?

3.11 Representing and solving word problems

1 One of these bar models is incorrect. Put a cross beside the incorrect representations. Complete the correct bar models.

a) Eilidh spent £1.25 on chocolate and £7.50 on a cinema ticket.

How much did she spend altogether?

£1.25	£7.50

b) Rhys spent two hours at the leisure centre. He played badminton for 45 minutes. How long did he spend in the rest of the centre?

120 minutes	
45 minutes	

c) Chung is 156cm tall. When he was five years old, he was 120 cm.

How much has he grown since he was five?

120 cm	156 cm

d) Skye has read 178 pages of her library book. She still has 189 pages to go.

How many pages does her book have?

178	189

2 Complete each Think Board.

Word problem	Bar model
Amari's gran was born in 1963. How old is she this year?	

Answer

Empty number line	Calculation

Word problem	Bar model

Bar model:

455	545

Answer

Empty number line	Calculation

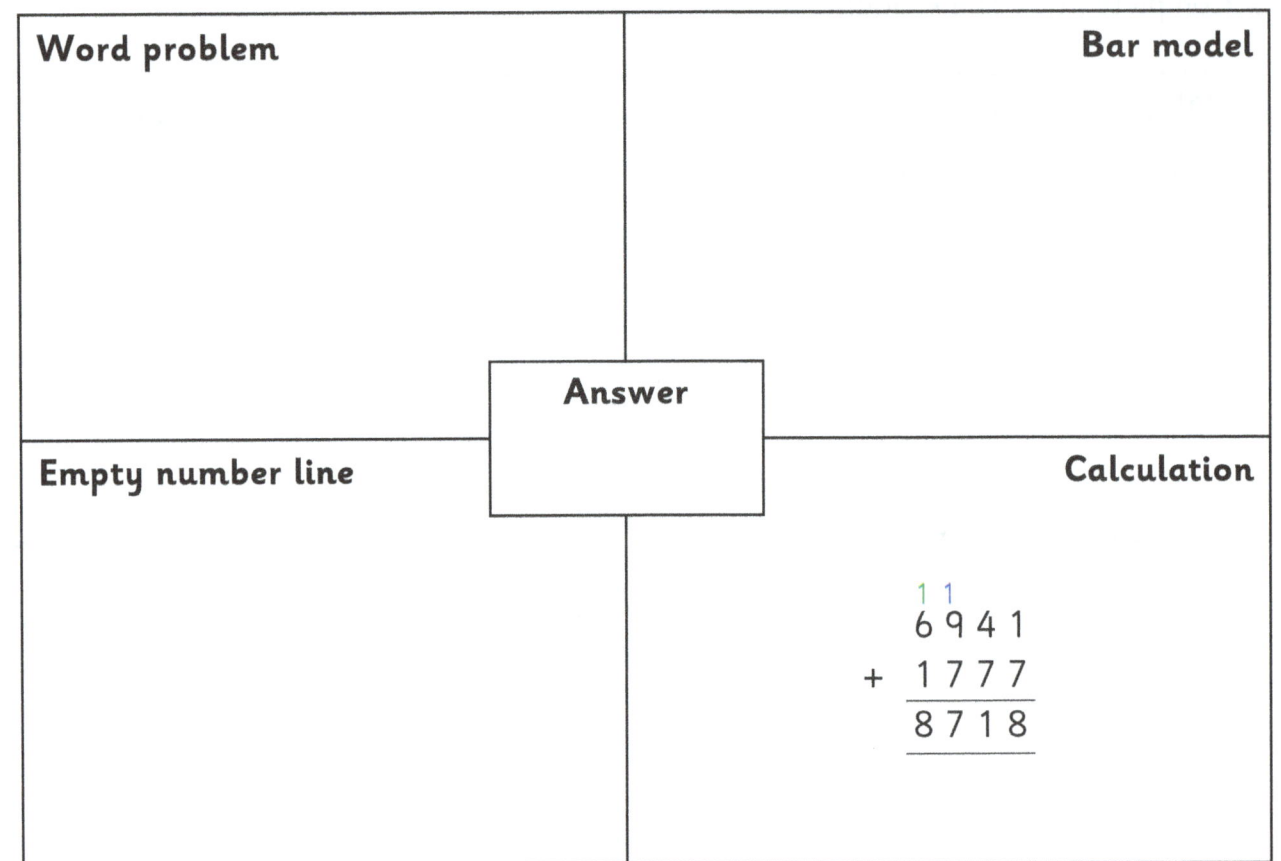

Word problem	Bar model

Answer

Empty number line	Calculation
	$\begin{array}{r} {\scriptstyle 1\ 1} \\ 6\,9\,4\,1 \\ +\ 1\,7\,7\,7 \\ \hline 8\,7\,1\,8 \end{array}$

⭐ **Challenge**

The local theatre has seats for 3059 people. At the beginning of June there are 1184 seats left for the Summer Show. By the middle of July another 543 tickets have been sold. How many free tickets are still available? Draw a bar model to show your thinking.

3.12 Solving multi-step word problems

1 The school fair hopes to raise £3000.

The raffle raises £1200 and the café raises half this amount.

How much more do they need to raise to have enough money?

2 The school library has 9897 books on the shelves.

The librarian puts 189 damaged books in the paper recycling bin.

The teachers get a donation of 165 books from the local book shop.

How many books does the library have now?

3 Mia starts a new job. She works every Monday, Tuesday, Thursday and Friday.

She is paid £1300 after her first month. She paid £275 for her train fares to work and £67 for lunches. She wants to buy a second-hand motorbike, which costs £1099. Is this possible? Explain your thinking.

4 The local football club has space for 8867 people. At the Saturday match there are 468 unsold tickets. There are 2041 children at the game.

How many adults are at the game?

5 Rory scored 1644 points playing his computer game.

Katie scored 2374 more than Rory and Sam beat both of them by scoring 709 more points than Katie.

How many points did Sam score?

6 Megan, her five-year-old sister and her mum and dad are looking forward to a family holiday.

The flights are £465 for adults and £325 for children.

The hotel is £1850 for a week.

How much will the trip cost?

7 Harry's uncle has walked 132 miles to raise money for charity.

He still has 248 miles to go until the finish point.

How far does he still need to walk before he is halfway there?

8 In January, the supermarket sells 8076 eggs. In February, it sells half that amount.

How many eggs are sold over the two months?

⭐ **Challenge**

On Monday the baker bakes 242 chocolate muffins, 164 cupcakes and 34 fruitcakes.

She sells three quarters of the goods in the bakery shop and donates a quarter to the local bowling club.

How many cakes does the bowling club get?

1 Round and adjust these calculations. The first one has been done for you.

a) 6·5 + 99 = 5·5 + 100

b) 8·5 + 199 =

c) 9·3 + 149 =

d) 149 + 19·3 =

e) 148 + 29·3 =

f) 31·9 + 5·4 =

g) 31·8 + 5·4 =

h) 31·5 + 5·7 =

2 Use partitions to solve these calculations. The first one has been done for you.

a) 62·3 + 3·6 = 60 + 2 + 3 + 0·3 + 0·6 = 65 + 0·9 = 65·9

b) 72·3 + 3·6 =

c) 72·4 + 4·6 =

d) 72·4 + 14·6 =

e) 78·4 + 14·8 =

f) 78·5 + 24·5 =

g) 88·5 + 24·6 =

3 Use a strategy of your choice to solve these calculations.

a) 8·8 + 4·4 =

b) 18·8 + 14·4 =

c) 20·4 + 4·9 =

d) 24·9 + 7·7 =

e) 24·8 + 6·7 =

f) 14·7 + 12·3 =

g) 14·3 + 12·7 =

4·7   13·9

12·6   7·4   4·9

12·8   6·5

Choose two numbers to create an addition where round and adjust is a good strategy. Solve your problem. Now choose two different numbers and solve by partitioning. You can use each number only once.

Round and adjust

Partitioning

1 Complete the table. The first one is done for you.

Decimal fractions < 1	Ones	Tenths	Number
0·6 + 0·6	1	2	1·2
0·6 + 0·8			1·4
0·8 + 0·7			
0·8 +	1	6	
+ 0·7	1	4	
0·9 +		8	

2 Use a number line to find the answers.

a) 23·5 + ☐ = 40

23·5   40

b) 23·7 + ☐ = 50

23·7   50

c) 33·7 + ☐ = 60

33·7   60

d) $60 = \boxed{} + 33 \cdot 1$

|——————————————————————————————————|
33·1   60

e) $60 = \boxed{} + 43 \cdot 2$

|——————————————————————————————————|
43·2   60

f) $73 \cdot 2 + \boxed{} = 100$

|——————————————————————————————————|
73·2   100

---

3 Find the missing number in each bar model.

a)

50	
13·9	

b)

55	
21·9	

c)

21·9	38·1

d)

46·6	34·4

★ Challenge

1. Complete the magic square using the numbers 0·2, 0·3, 0·5. 0·7 and 0·8. Each row, column and diagonal must have the same total.

		0·6
0·4	0·9	

2. Experiment with making your own magic square with decimal fractions!

1 Double these numbers:

a) Double 6 = [ ]

b) Double 9 = [ ]

c) Double 12 = [ ]

d) Double 15 = [ ]

e) Double 18 = [ ]

f) Double 21 = [ ]

g) Double 24 = [ ]

h) Double 27 = [ ]

i) Double 30 = [ ]

j) Double 33 = [ ]

2 Calculate these by using your knowledge of the 3 times table and doubling. The first one has been done for you.

a) $3 \times 2 =$ [6]

b) $3 \times 4 =$ [ ]

c) $3 \times 6 =$ [ ]

$6 \times 2 =$ [12]

$6 \times 4 =$ [ ]

$6 \times 6 =$ [ ]

d) $3 \times 8 =$ [ ]

e) $3 \times 10 =$ [ ]

f) $3 \times 5 =$ [ ]

$6 \times 8 =$ [ ]

$6 \times 10 =$ [ ]

$6 \times 5 =$ [ ]

3 Complete the fact families. One has been done for you.

a)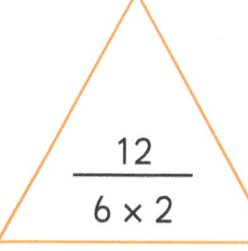

b)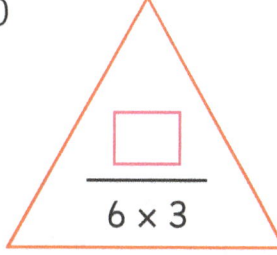

c)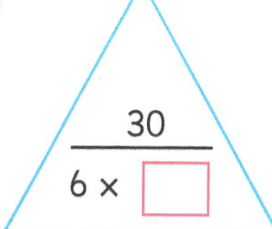

d)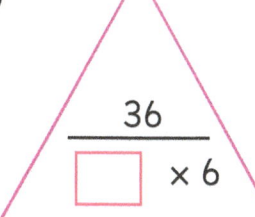

e)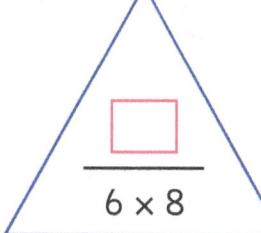

f)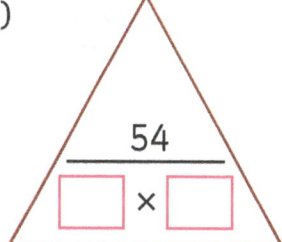

Write the fact families for 6 up to 6 × 10.

$6 \times 1 =$
$1 \times 6 =$
$6 \div 1 =$
$6 \div 6 =$

1 Calculate these by using your knowledge of the 10 times table. The first one has been done for you.

a)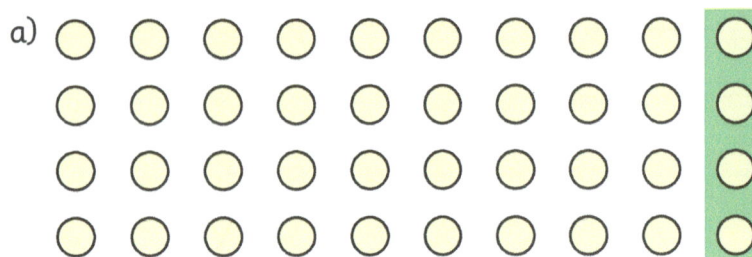

$4 \times 9 = (4 \times 10) - (4 \times 1)$

$= 40 - 4$

$= 36$

b)

$8 \times 9 = ($ ⬚ $\times 10) - ($ ⬚ $\times 1)$

$=$ ⬚ $-$ ⬚

$=$ ⬚

c)

$7 \times 9 = ($ ⬜ $\times 10) - ($ ⬜ $\times 1)$

$= $ ⬜ $-$ ⬜

$= $ ⬜

---

2 Complete the following:

a) $9 \times 10 = $ ⬜

b) $45 \div 9 = $ ⬜

c) ⬜ $\div 9 = 2$

d) $4 \times 9 = $ ⬜

e) $54 \div 9 = $ ⬜

f) ⬜ $\div 9 = 8$

g) ⬜ $\div 9 = 7$

h) $3 \times 9 = $ ⬜

i) $99 \div 9 = $ ⬜

j) ⬜ $\div 9 = 1$

1. Complete the table.

	Tens	Ones
9 × 1	0	9
9 × 2		
9 × 3		
9 × 4		
9 × 5		
9 × 6		

2. Look at the pattern of your answers. What do you notice? Keep exploring – what is the first fact when this does not happen?

Use this box to explore the pattern.

1 Circle the facts that are incorrect and correct them. Now fill in the missing facts.

	× 10	× 100	× 1000
65		6500	
70	7000		
75			75 000
80			8000
85		850	
90	900		
95		9500	

2 Complete these calculations. The first one is done for you.

a) 4 × 30 = 4 × 3 tens

      = 12 tens

      = 120

b) 4 × 40 = 4 × ☐ tens

      = ☐ tens

      = ☐

c) 4 × 400 = 4 × ☐ hundreds

      = ☐ hundreds

      = ☐

d) 5 × 600 = 5 × ☐ hundreds

      = ☐ hundreds

      = ☐

3 Calculate the following:

a) 9×7 = [ ]

b) 6×8 = [ ]

90×7 = [ ]

60×8 = [ ]

900×7 = [ ]

6000×8 = [ ]

9000×7 = [ ]

600×8 = [ ]

c) Glendale Football Club charges £8 for a child ticket to see a match. They sell 600 child tickets. How much money do the make?

d) An adult ticket to see a match is £40. Ava's mum buys tickets for herself and six friends. How much does she pay?

e) The shirt factory uses seven buttons on each shirt. How many buttons do they need to make 4000 shirts?

Use your knowledge of multiplying by multiples of 10, 100 and 1000 to solve these.

1. The answer is 2400. What could the question be? How many different ways can you find?

2. The answer is 12 000. What could the question be? How many different ways can you find?

3. The answer is 420. What could the question be? How many different ways can you find?

4.4 Dividing by multiples of 10, 100 or 1000

1 Complete the table.

	÷ 10	÷ 100	÷ 1000
6000			
8000			
2000			
5000			

2 Complete these calculations. The first one is done for you.

a) $160 ÷ 8 = 16$ tens $÷ 8$

        $= 2$ tens

        $= 20$

b) $240 ÷ 8 = \boxed{}$ tens $÷ 8$

        $= \boxed{}$ tens

        $= \boxed{}$

c) $3200 ÷ 8 = \boxed{}$ hundreds $÷ 8$

        $= \boxed{}$ hundreds

        $= \boxed{}$

d) $3200 ÷ 4 = \boxed{}$ hundreds $÷ 4$

        $= \boxed{}$ hundreds

        $= \boxed{}$

e) $32\,000 \div 4 =$ ☐ thousands $\div 4$ f) $36\,000 \div 4 =$ ☐ thousands $\div 4$

$=$ ☐ thousands $=$ ☐ thousands

$=$ ☐ $=$ ☐

3 Use these known facts to help you.

| $49 \div 7 = 7$ | $36 \div 6 = 6$ | $25 \div 5 = 5$ | $16 \div 4 = 4$ |

a) $4900 \div 7 =$ ☐   b) $250 \div 5 =$ ☐   c) $36\,000 \div 6 =$ ☐

d) $1600 \div 4 =$ ☐   e) $160 \div 4 =$ ☐   f) $3600 \div 6 =$ ☐

g) $490 \div 7 =$ ☐   h) $25\,000 \div 5 =$ ☐   i) $2500 \div 5 =$ ☐

4 Solve the following:

a) Apples are sold in bags of five. The school cook orders 4500 apples.
   How many bags does she order?

b) Bananas are sold in bunches of six. The school cook orders 540 bananas.
   How many bunches does she order?

c) Oranges are sold in bags of four. The school cook orders 3600 oranges.
   How many bags does she order?

Fill in the blanks using the digits 0, 2 and 4. You can use each digit as often as you like. All numbers must be multiples of 10, 100 or 1000.

☐☐☐ ÷ ☐☐ = ☐

☐☐☐☐ ÷ ☐☐ = ☐

☐☐☐☐☐ ÷ ☐☐ = ☐

4.5 Solving division problems

1

Solve the following calcuations. Use the array to help you.

a) $3 \times 5 = $ ▢

b) ▢ $\times 3 = 15$

c) ▢ $\div 5 = 3$

d) $15 \div$ ▢ $= 5$

2 Complete the following calculations. The first has been done for you.

a) What is 18 divided by 2? ▢ 9 ▢ $\times 2 = 18$ so $18 \div 2 = $ ▢ 9 ▢

b) What is 18 divided by 3? ▢ $\times 3 = 18$ so $18 \div 3 = $ ▢

c) What is 18 divided by 9? ▢ $\times 9 = 18$ so $18 \div 9 = $ ▢

d) What is 36 divided by 9? ▢ $\times 9 = 36$ so $36 \div 9 = $ ▢

e) What is 36 divided by 6? ▢ $\times 6 = 36$ so $36 \div 6 = $ ▢

f) What is 36 divided by 4? ▢ $\times 4 = 36$ so $36 \div 4 = $ ▢

g) What is 36 divided by 3? ▢ $\times 3 = 36$ so $36 \div 3 = $ ▢

3 Complete the following by writing x or ÷ in each box. The first has been done for you.

a) 12 ☐× 2 = 24

b) 24 ☐ 2 = 12

c) 24 ☐ 12 = 2

d) 2 ☐ 12 = 24

e) 24 ☐ 3 = 8

f) 8 ☐ 3 = 24

g) 4 ☐ 6 = 24

h) 24 ☐ 6 = 4

4 Use the inverse relationship between multiplication and division to solve these.

a) £50 is shared between two brothers.
How much does each brother get?

b) 50 pencils are shared between ten packets.
How many go in each packet?

c) 100 books are shared equally between ten classes.
How many does each class get?

d) 25 tennis balls are shared equally between five groups.
How many does each group get?

e) 45 lemons are shared equally between five school cooks.
How many does each cook get?

f) £45 is shared equally between nine children.
How much does each child get?

1) Use these facts to answer the following:

| 16 × 7 = 112 | 362 × 14 = 5068 | 452 ÷ 113 = 4 |

a) 112 ÷ 7 = ⬚

b) 14 × ⬚ = 5068

c) 452 ÷ ⬚ = 113

d) 7 × 16 = ⬚

e) 5068 ÷ 362 = ⬚

f) 14 × ⬚ = 5068

g) 112 ÷ 16 = ⬚

h) 4 × ⬚ = 452

2) Now create three calculations using the facts above.

4.6 Solving multiplication problems

1 Partition these numbers by place value. The first one is one for you.

a) 36 = 30 + 6

b) 26 = ☐

c) 28 = ☐

d) 58 = ☐

e) 158 = ☐

f) 142 = ☐

2 Use the grids to answer the following:

a) 132 × 4

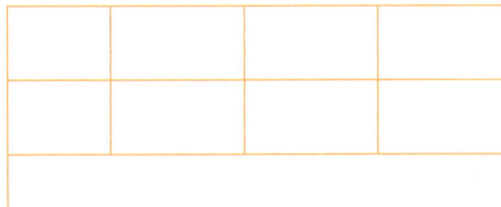

b) 138 × 4

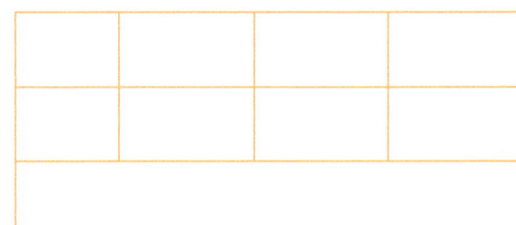

c) 138 × 5

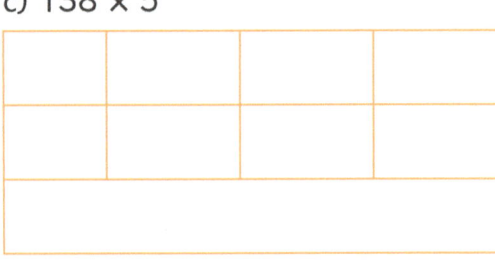

d) 238 × 6

3 Find the answers. Draw grids to show your calculations.

a) Mr Green earned £424 per week for four weeks. How much did he earn during that time?

b) Ms Lemon got a delivery of sweets for her shop. She got eight boxes with 164 chocolate bars in each. How many bars of chocolate did she get?

c) Asha saves £8.45 each week for nine weeks. How much did she have in it at the end of nine weeks?

★ **Challenge**

- Make three different 3-digit numbers using the digits 3, 6 and 9.
- Ask a partner to choose a number between 2 and 9 for you to multiply your numbers by.

Use multiplication grids to help you find the answers.

1.

2.

3.

4.7 Using partitioning to solve division problems

1 Match the number to the correct partition.

56   60 + 21

81   33 + 33

66   60 + 33

48   40 + 16

93   24 + 24

2 Use the partitions in question 1 to complete these division calculations. Complete the grids. The first one is done for you.

a) 56 ÷ 4

÷	40	16
4	10	4
10 + 4 = 14		

b) 81 ÷ 3

÷	60	21
3		
+ =		

c) 66 ÷ 3

÷	33	33
3		
+ =		

d) 48 ÷ 6

÷		
6		
+ =		

e) 93 ÷ 3

÷		
+ =		

3 Use the partition method to solve these problems.

There are 132 students in S2. How many groups will be needed if there are:

a) three students in each group?

b) four students in each group?

c) six students in each group?

★ Challenge

1. Use different partitions to show 168 ÷ 4.

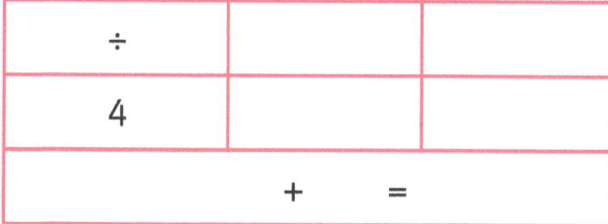

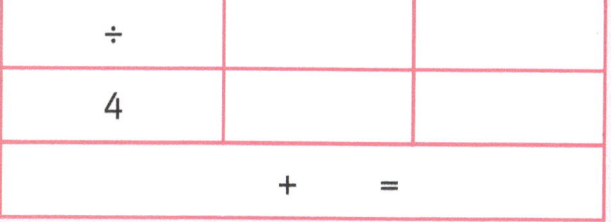

2. Which partition did you find worked best? Why might that be?

4.8 Using rounding and compensating to solve multiplication problems

1 a) Use the diagram to show how 4 × 20 can help you answer 4 × 19.

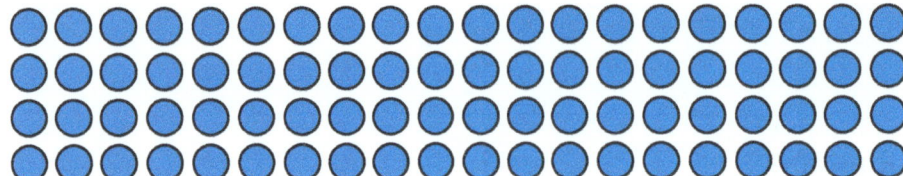

4 × 19 = _____

b) Use the ten frames to show 9 × 9 using compensation.

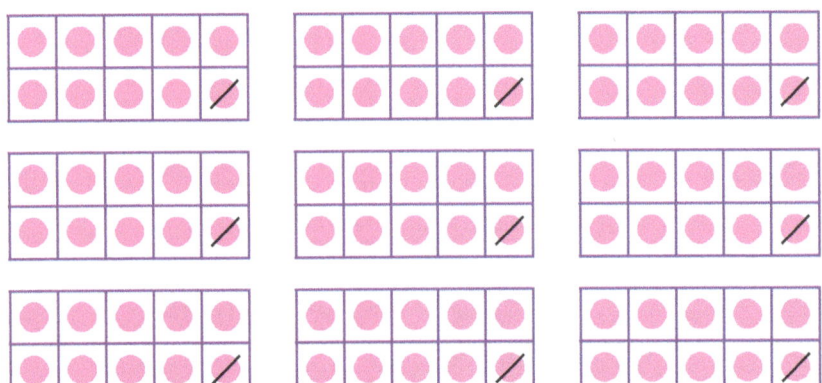

9 × 10 = _____   _____ − 9 = 81

So 9 × 9 = _____

c) Draw ten frames or use a set of ten frames to help you solve these:

i) 3 × 9

ii) 11 × 9

2 Use the number line to show:

a) 19 × 4 = ☐

|———————————————————————————————
0

b) 18 × 4 = ☐

|———————————————————————————————
0

3 Use rounding and compensation to help you solve these problems:

a) 5 × 40 = 200 so 5 × 39 = ☐   5 × 41 = ☐

b) 6 × 40 = 240 so 6 × 39 = ☐   6 × 38 = ☐

★ Challenge

0 1 2 3 4 5

Leo used two of these digits to create a number. He multiplied his two-digit number by one of the remaining digits and got an answer of 120.

1. What was Leo's two-digit number?

2. What did he multiply his two-digit number by?

4.9 Solving problems involving addition, subtraction, multiplication and division

1   Ryan says that he can solve 36×4 by adding.
    Dylan says that is not possible. Who is correct and why?

2   Edinburgh is 648km from London, Cardiff is 240km from London and Belfast is 742km from London. Mr Wood and Ms Stone are lorry drivers.

    a) Mr Wood drives from Edinburgh to London and back again twice in July. How far does he travel?

    b) Ms Stone drives from Cardiff to London and back three times in July. Who drove further in July and by how much?

3 The supermarket has a deal on fruit. Lemons, oranges and limes are £2 for a bag of six or they can be bought for 35p each. Apples cost 40p each or £3 for a bag of six.

a) Andi's mum buys 36 lemons. What is the least she has to pay?

b) Andi's dad buys 12 apples. What is the least he has to pay?

c) Andi's grandad buys three bags of limes and three bags of apples. How much does he pay?

4 Four students are selling cookies they have made. They sell them for 50p each. How much will they each get if they sell:

a) 124 cookies at school?

b) 96 cookies at the local library?

c) 280 cookies at the village show?

d) How much money do the students make altogether?

5 Mr Young has been offered two jobs.

- The pet shop pays £12 each hour and he would work 36 hours each week.
- The garage pays £14 each hour and he would work 32 hours each week.

Which job would pay the most?

⭐ **Challenge**

Roisin's 10th birthday treat is a visit to the cinema. She is going with her mum, her aunt, her uncle and her three cousins.

What is the cheapest way they can get tickets?

Remember to show your working.

CINEMA

01234

Adult: £11
Child: £7
Family: £30 (2 adults, 2 children)

★ ★ ★ ★ ★

01234

4.10 Multiplying decimal fractions

1 Complete these calculations. The first one is done for you.

a) [4] × 3 = 12

b) [ ] × 3 = 120

c) [ ] × 3 = 1200

d) [ ] × 3 = 12 000

e) [ ] × 6 = 24

f) [ ] × 6 = 240

g) 400 × [ ] = 2400

h) 4000 × [ ] = 24 000

2 Record your answers in the table. The first one has been done for you.

	Thousands	Hundreds	Tens	Ones	•	Tenths
a) 0·6				0	•	6
b) 0·6 × 10					•	
c) 0·6 × 100					•	
d) 0·6 × 1000					•	

3 Multiply these decimals and record the answer on the grid.

a)

Tens	Ones	Decimal point	Tenths
	0	.	1

 × 10

Tens	Ones	Decimal point	Tenths
		.	

b)

Tens	Ones	Decimal point	Tenths
	0	.	3

 × 10

Tens	Ones	Decimal point	Tenths
		.	

c)

Tens	Ones	Decimal point	Tenths
	1	.	3

 × 10

Tens	Ones	Decimal point	Tenths
		.	

d)

Tens	Ones	Decimal point	Tenths
	2	.	3

 × 10

Tens	Ones	Decimal point	Tenths
		.	

4 Multiply these decimals and record the answer on the grid.

a)

Tens	Ones	Decimal point	Tenths
	0	.	1

× 100

Tens	Ones	Decimal point	Tenths
		.	

b)

Tens	Ones	Decimal point	Tenths
	0	.	3

× 100

Tens	Ones	Decimal point	Tenths
		.	

c)

Tens	Ones	Decimal point	Tenths
	1	.	3

× 100

Tens	Ones	Decimal point	Tenths
		.	

d)

Tens	Ones	Decimal point	Tenths
	2	.	3

× 100

Tens	Ones	Decimal point	Tenths
		.	

5 Multiply these decimals.

a) 1·7 × 10 = [　　　]

b) 1·7 × 100 = [　　　]

c) 1·7 × 1000 = [　　　]

d) 3·7 × 10 = [　　　]

e) 4·7 × [　　　] = 470

f) 5·7 × [　　　] = 5700

g) [　　　] × 10 = 53

h) [　　　] × 100 = 53

6 There are 1000g in a kg. How many grams are there in:

a) 2kg = [　　　]

b) 2·5kg = [　　　]

c) 2·25kg = [　　　]

d) 2·75kg = [　　　]

e) 0·75kg = [　　　]

f) 0·66kg = [　　　]

g) 5·66kg = [　　　]

h) 12·66kg = [　　　]

⭐ **Challenge**

The answer is 5700. What might the question be?
Using just the cards, show as many questions as possible.

[　　　　　　　　　　]

(0·57)　　(57)　　(0·057)

(570)　　(5·7)　　(5700)

× 10　　× 100　　× 1000

4.11 Dividing whole numbers by 10, 100 and 1000

1 Decide whether these calculations are correct or incorrect. Use ✔ or ✖.

a) 720 ÷ 10 = 72

b) 7200 ÷ 100 = 720

c) 72 000 ÷ 1000 = 72

d) 72 000 ÷ 10 = 720

2 Divide these numbers and record the answers on the grids.

a)

Hundreds	Tens	Ones
	4	5

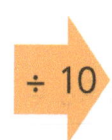 ÷ 10

Tens	Ones	Decimal point	Tenths
		.	

b)

Hundreds	Tens	Ones
4	5	0

÷ 10

Tens	Ones	Decimal point	Tenths
		.	

c)

Hundreds	Tens	Ones
4	5	5

÷ 10

Tens	Ones	Decimal point	Tenths
		.	

d)

Hundreds	Tens	Ones
4	0	5

÷ 10

Tens	Ones	Decimal point	Tenths
		.	

3 Divide these numbers and record the answers on the grids.

a)

Thousands	Hundreds	Tens	Ones
1	7	6	0

÷ 100

Tens	Ones	Decimal point	Tenths
		.	

b)

Thousands	Hundreds	Tens	Ones	÷ 1000	Tens	Ones	Decimal point	Tenths
1	7	1	0				.	

c)

Thousands	Hundreds	Tens	Ones	÷ 1000	Tens	Ones	Decimal point	Tenths
1	7	6	0				.	

d)

Thousands	Hundreds	Tens	Ones	÷ 100	Tens	Ones	Decimal point	Tenths
		1	7				.	

---

4 Divide these numbers.

a) $96 \div 10 =$ ☐

b) $96 \div 100 =$ ☐

c) $9 \div 100 =$ ☐

d) $90 \div 1000 =$ ☐

e) $900 \div$ ☐ $= 0.9$

f) $9 \div$ ☐ $= 0.9$

g) ☐ $\div 10 = 0.7$

h) ☐ $\div 100 = 0.7$

---

5 There are 100cm in a metre. Convert these cm to m.

a) 560cm = ☐

b) 56cm = ☐

c) 5600cm = ☐

d) 725cm = ☐

e) 775cm = ☐

f) 750cm = ☐

1. Using the cards shown below, create ten different number sentences.

7200   7·2   720   ÷ 10

7200   0·72   ÷ 100

÷ 1000

2. Now find the answers to the questions you created.

For example:

7200 ÷ 100 = 72

1 Order these numbers from smallest to biggest.

0·3 $\frac{4}{8}$ 5·2   0·25

2 A P.E. teacher sorts 30 students into four teams for a game of football.

How many students are left without a team?

3 The dinner tables in the school hall each sit six people.

How many tables will be needed so that 64 students can each have a seat?

4 Iga has a pack of cards. She deals 38 cards between eight players.

How many cards are left over?

5. Granny gives Willow and her sister £13 to share.

How much do they each get?

6. Ms Brown goes to the café with her three friends. The bill comes to £26 and they all pay the same amount.

How much does each person pay?

7. Ethan is saving up for a new game. It costs £32·50 and he saves the same amount each week. After five weeks he has saved exactly £32·50.

How much did he save each week?

8. Mrs Black bakes three cakes which she shares between her two sisters.

How many cakes does each sister get?

9 Mrs Wood bakes 21 muffins which she shares between her six children.

How many muffins does each child get?

10 Cole takes the bus to see his Gran. The bus travels six miles and takes 45 minutes.

How long does it take to travel one mile?

⭐ **Challenge**

Which question best matches which answer?   **5·75**   **5 $\frac{3}{4}$**   **5 remainder 3**

1. Bob has baked 23 cookies which he sells in bags of four. How many bags will he fill?

2. Bob spent £23 on four copies of a book for himself and his three friends. How much was each book?

3. Bob ran four races over the weekend. It took him 23 hours in total and each race took him the same length of time. How long did each race take?

4.13 Solving multiplication and division problems

Complete these Think Boards.

1

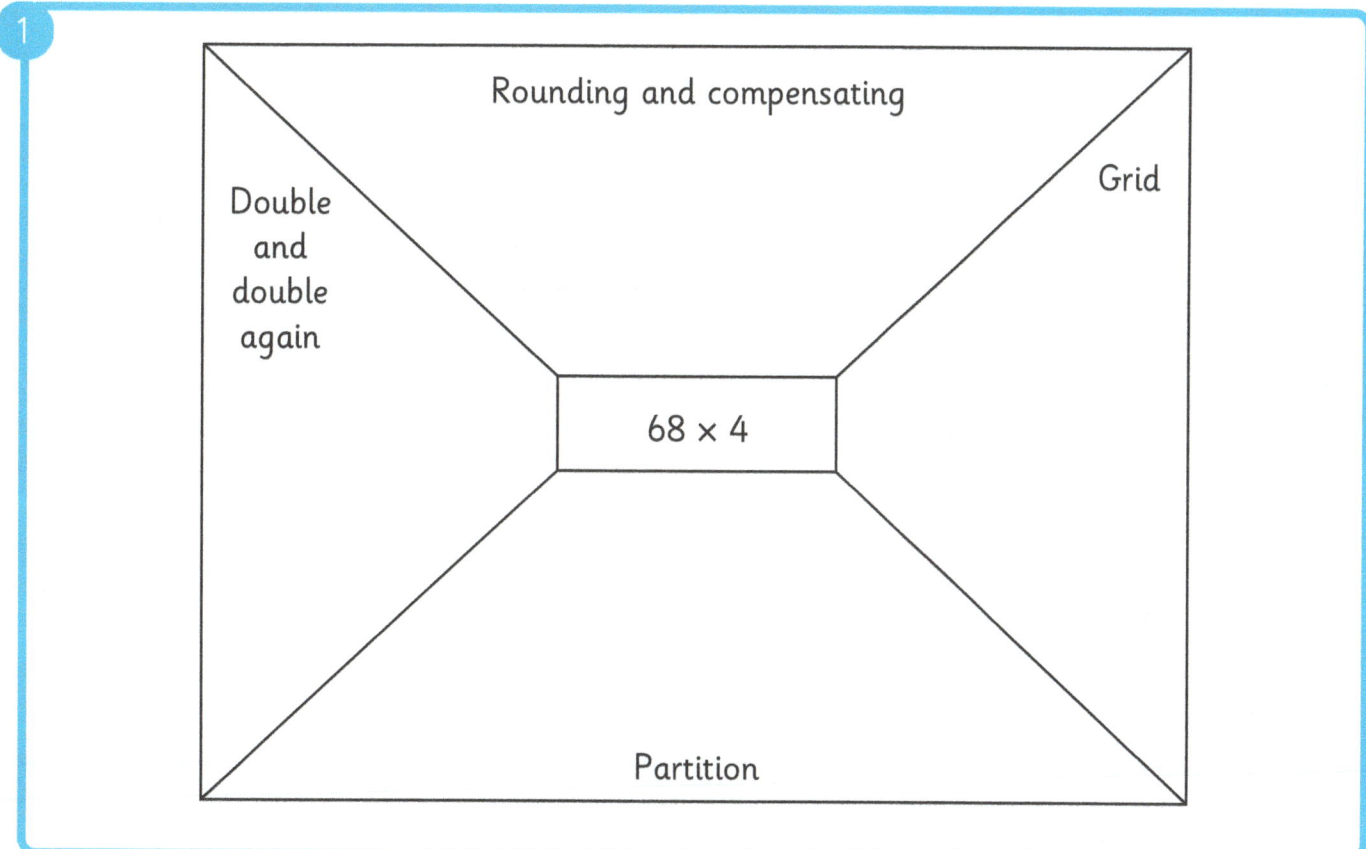

Rounding and compensating

Double and double again

Grid

68 × 4

Partition

2

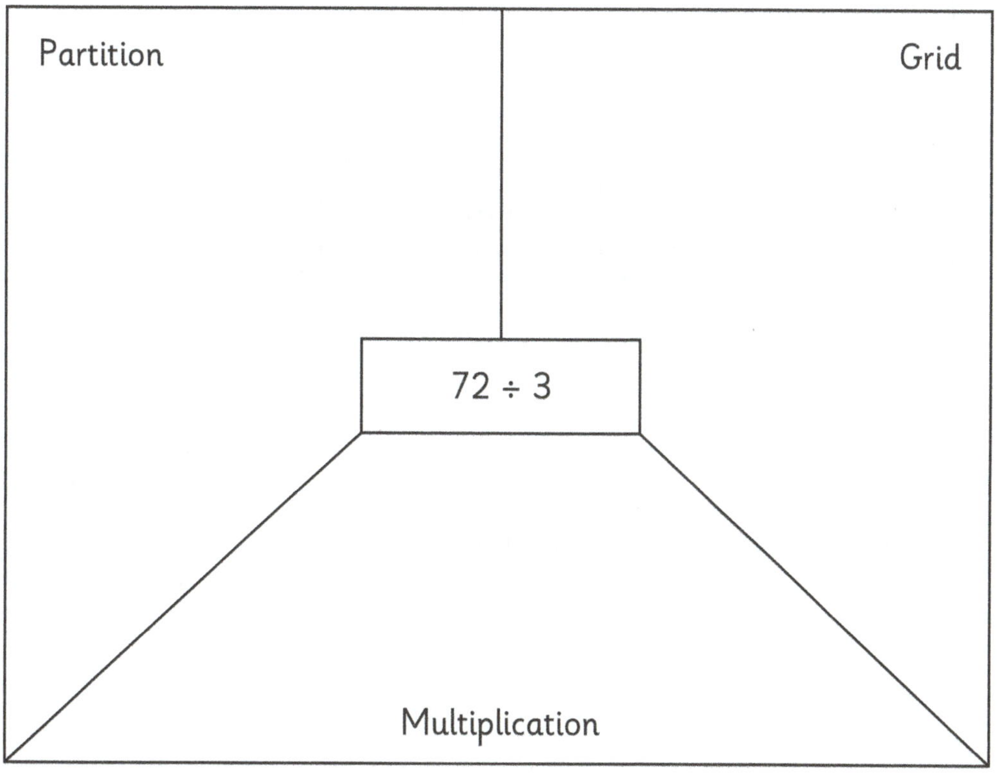

Partition

Grid

72 ÷ 3

Multiplication

3

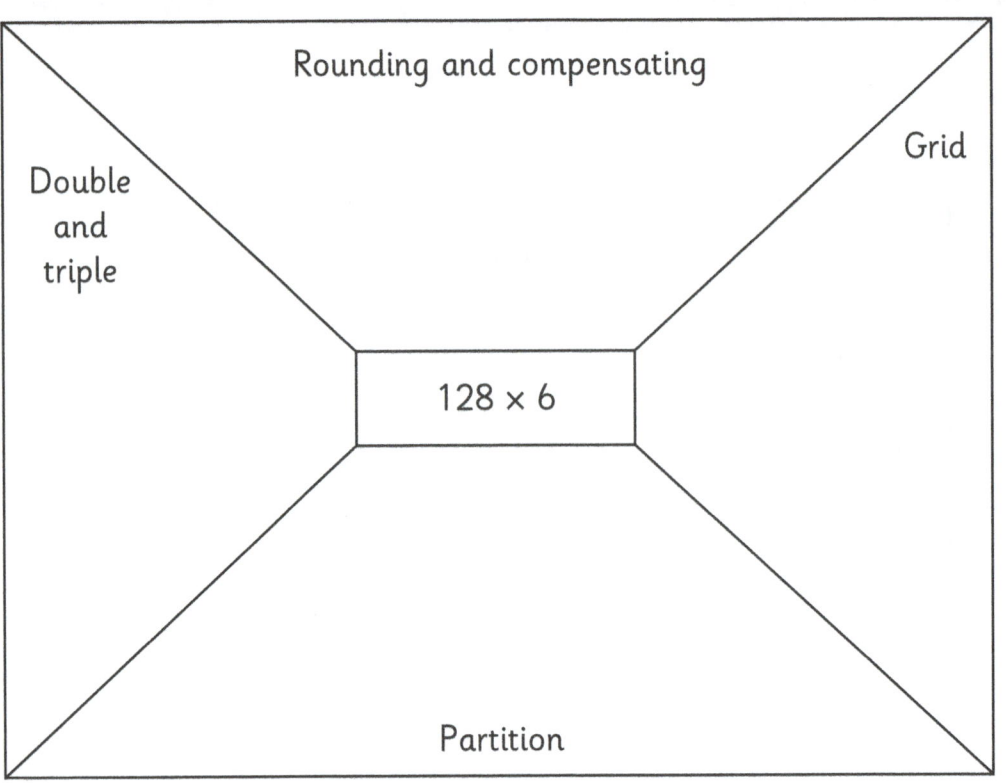

Rounding and compensating

Grid

Double
and
triple

128 × 6

Partition

4

Partition

Grid

114 ÷ 6

Multiplication

Emily has a great magic trick which means she ALWAYS knows the answer. Dan doesn't believe Emily's trick will always work. What do you think?

Try Emily's trick for different numbers. Compare your findings with a friend.

a) Write down any number.

b) Add 5 to your number.

c) Multiply this new number by 3.

d) Now subtract 15.

e) Divide your answer by the number you first thought of.

f) Next add 7.

g) The answer is ...

Attempt 1

Attempt 2

Attempt 3

Attempt 4

Attempt 5

Who is correct?

1 Complete the Venn diagram.

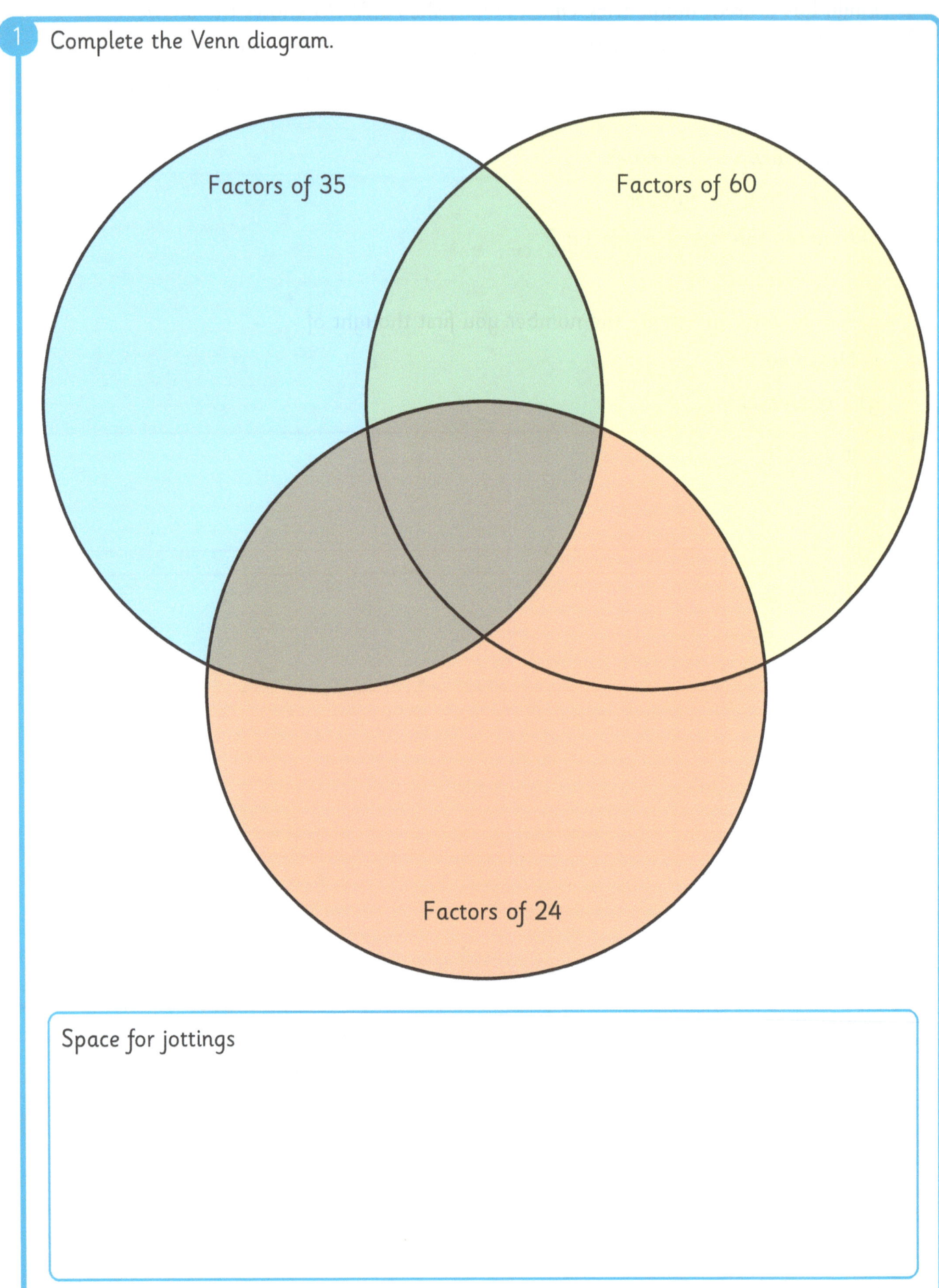

Factors of 35

Factors of 60

Factors of 24

Space for jottings

2 Use the numbers 1 to 30. You can use each number only once. Find a number that has:

a) an odd number as a factor

b) a factor greater than 14

c) an even number as a factor

d) only three factors

e) a factor that is less than 3

f) only two factors

3 a) Roll a dice. Which numbers less than 35 have your number as a factor?

Repeat this activity five times.

b) Use two dice and total them. Which numbers less than 110 have your number as a factor? Repeat this activity five times.

⭐ Challenge

I am a number between 1 and 45 and I have only eight factors.

Which number might I be?

1

1	2	3	4	5	6	7	8	9	10
11	12	13	14	15	16	17	18	19	20
21	22	23	24	25	26	27	28	29	30
31	32	33	34	35	36	37	38	39	40
41	42	43	44	45	46	47	48	49	50
51	52	53	54	55	56	57	58	59	60
61	62	63	64	65	66	67	68	69	70
71	72	73	74	75	76	77	78	79	80
81	82	83	84	85	86	87	88	89	90
91	92	93	94	95	96	97	98	99	100

a) Colour all the multiples of 10 yellow.

b) Colour all the multiples of 3 blue.

c) Colour all the multiples of 7 red.

d) Which numbers have you coloured in yellow and blue?

e) Which numbers have you coloured in blue and red?

f) Which number have you coloured in yellow and red?

2 Write the numbers in the circles in the correct place in the diagram.

12　17

25　6

40　21

15　16

9　330

	Multiple of 5	Not multiple of 5
Multiple of 2		
Not multiple of 2		

3

	Multiple of 6	Multiple of 9	Neither a factor nor a multiple of 6 or 9
81		✔	
18			
108			
80			
1			
9			
91			
24			
48			

Find the lowest common multiple of these pairs. Have a multiplication square handy!

	Lowest Common Multiple	
3		4
3		5
3		6
6		9
5		9
4		9
3		9

Multiples of 3

Multiples of 4

Multiples of 5

Multiples of 6

Multiples of 9

6.1 Identifying equivalent fractions

1 Colour the tenths bar to show the equivalent fraction.

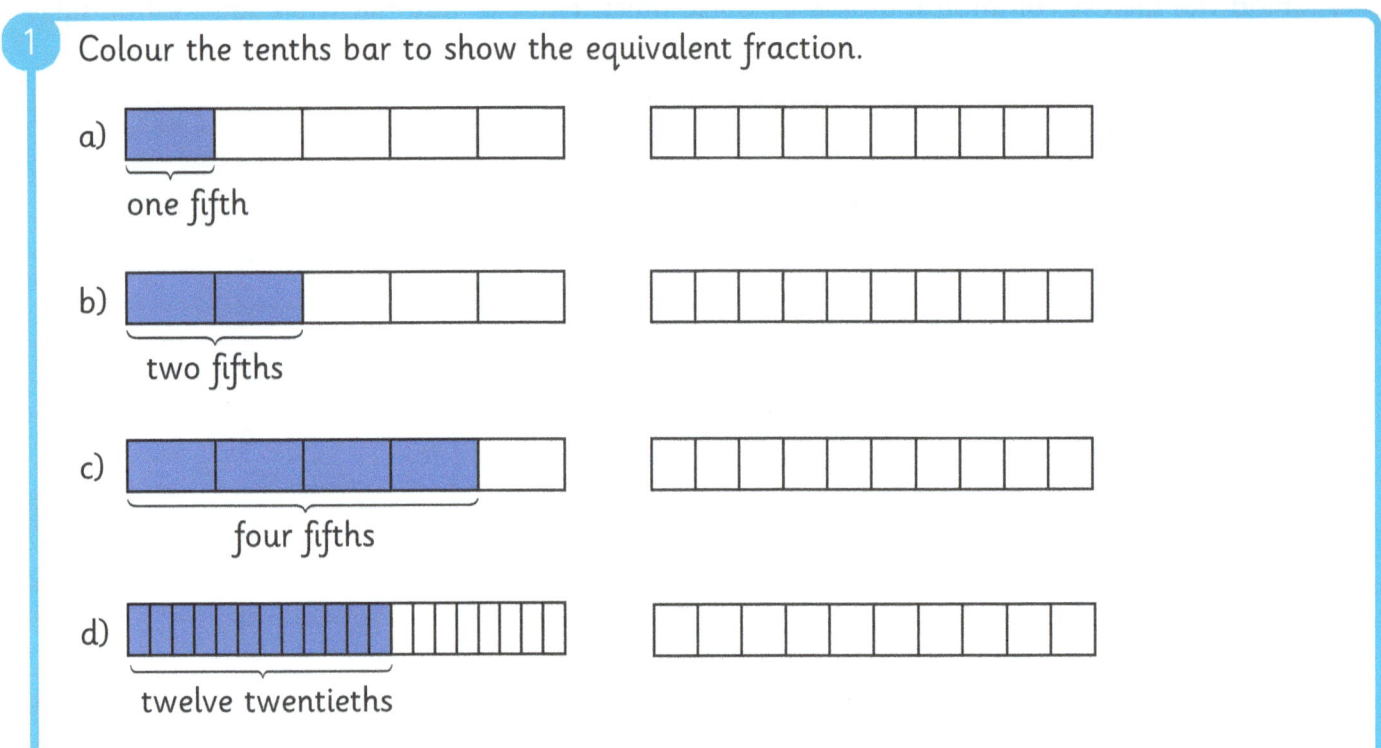

a) one fifth

b) two fifths

c) four fifths

d) twelve twentieths

2 Colour the hundredths bar to show the equivalent fraction.

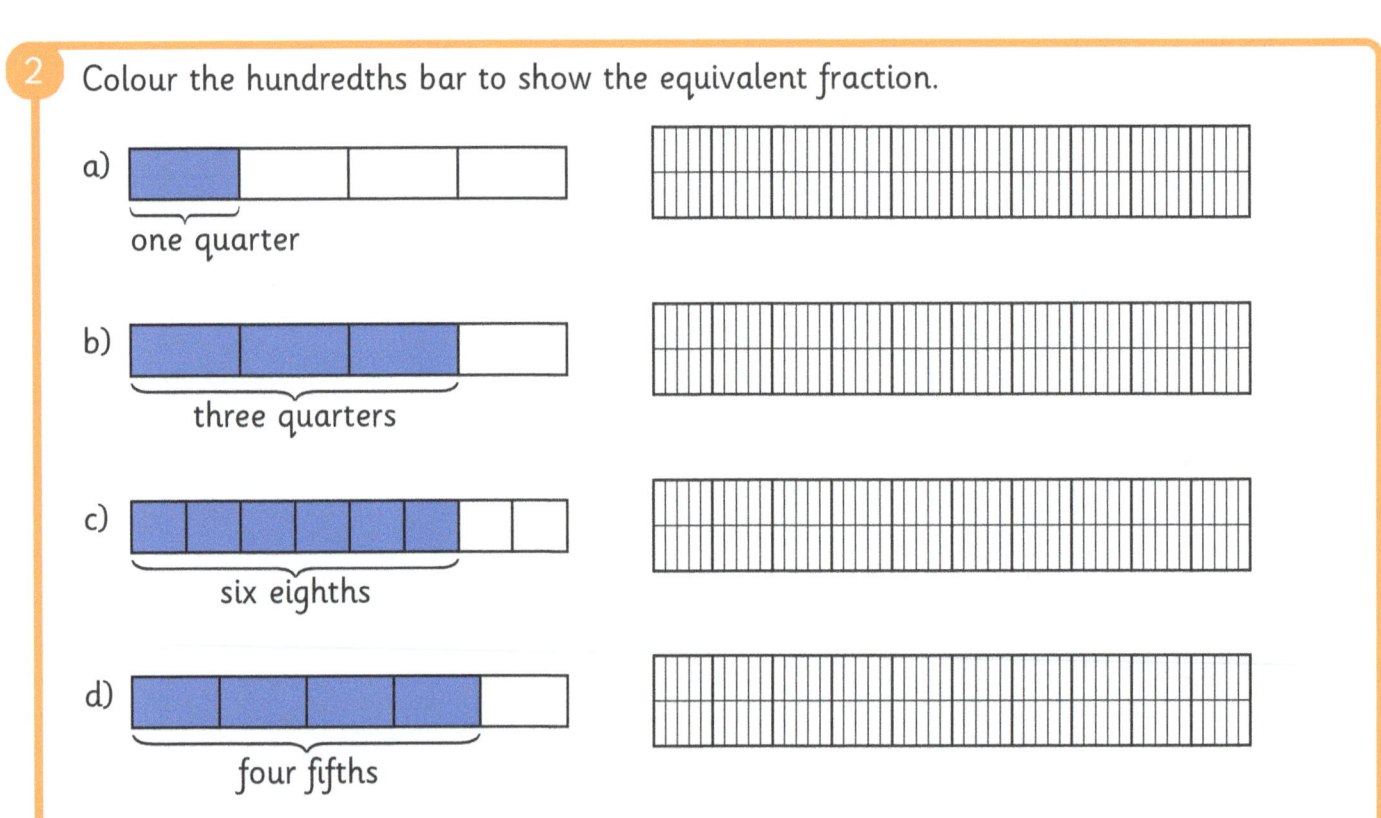

a) one quarter

b) three quarters

c) six eighths

d) four fifths

3 Which of these can be changed into tenths? Tick the fractions that can be changed.

a)

one half

b)

three fifths

c)

four
twentieths

⭐ **Challenge**

Colour each bar to create pairs of equivalent fractions. You may have to use your knowledge of simplifying. The first one is done for you.

a)

eight tenths   sixteen twentieths

b)

c)

d)

e)

6.2 Calculating equivalent fractions

1 Use doubling to find and write equivalent fractions to these.

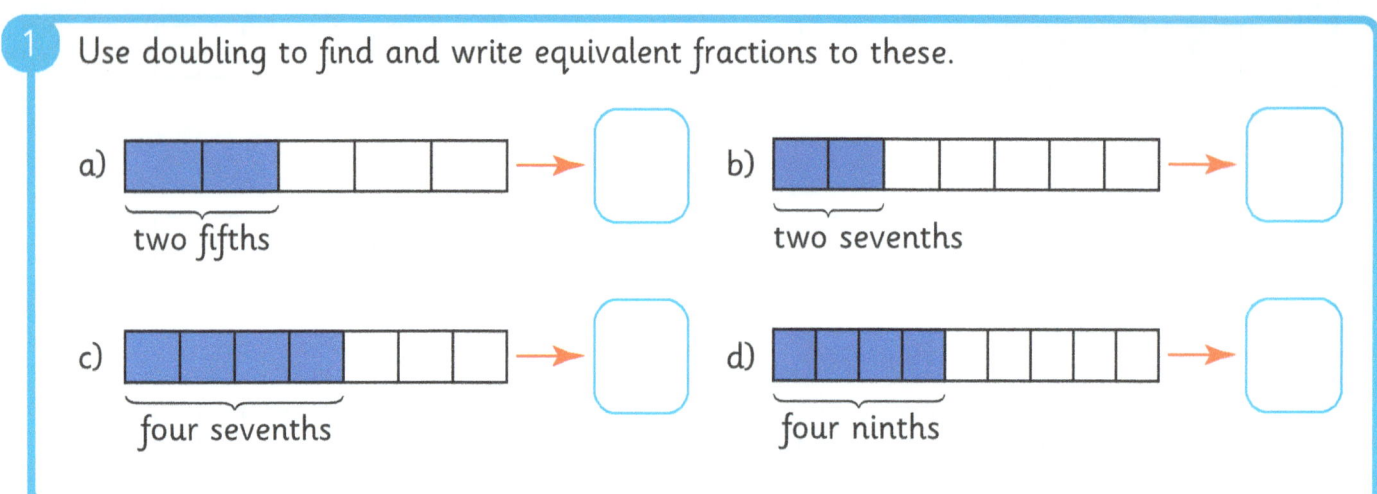

a)

two fifths

b)

two sevenths

c)

four sevenths

d)

four ninths

2 Use multiplication to find three equivalent fractions for each of these.

a)

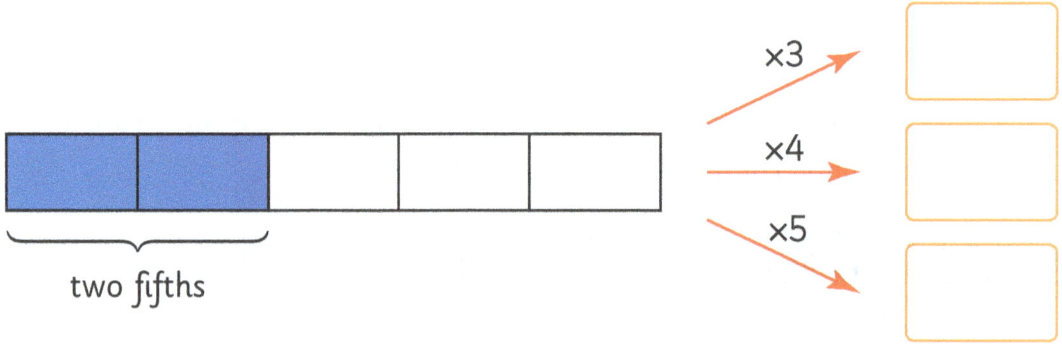

two fifths

×3

×4

×5

b)

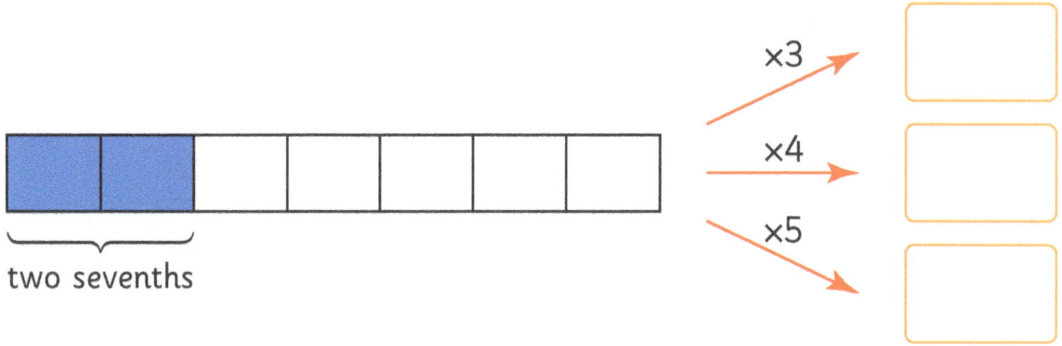

two sevenths

×3

×4

×5

c)

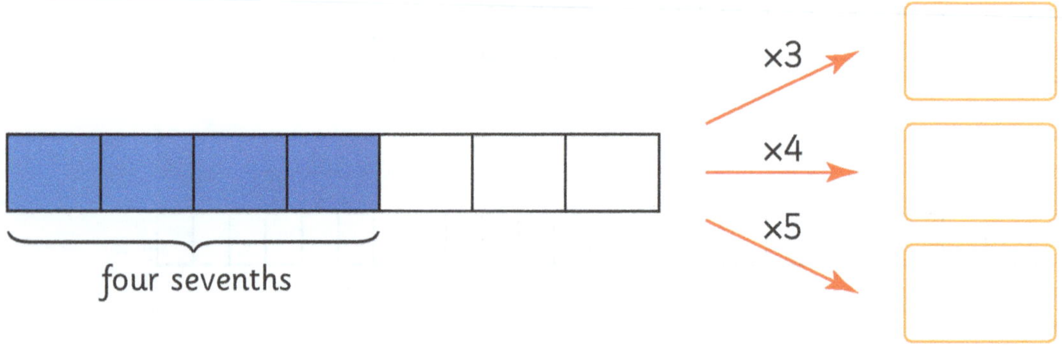

four sevenths

×3

×4

×5

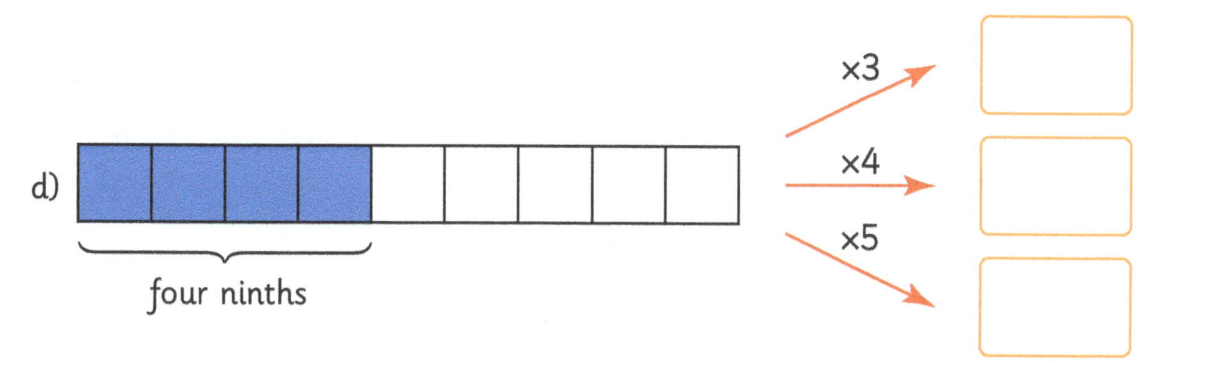

d) four ninths

×3
×4
×5

---

3 Use a multiplication of your own choice to find an equivalent fraction for each of these.

a)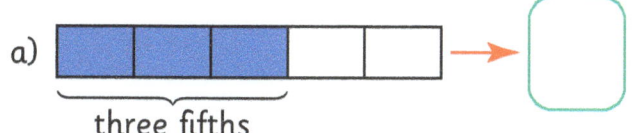
three fifths

b)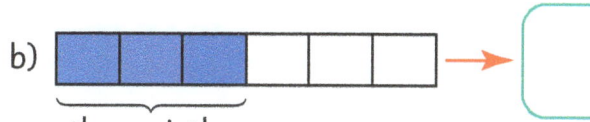
three sixths

c)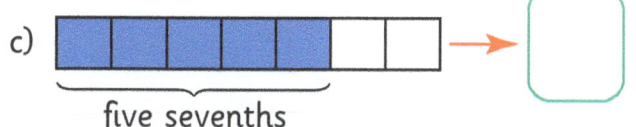
five sevenths

d)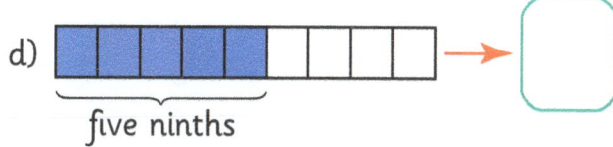
five ninths

⭐ **Challenge**

Use these digits to make equivalent fractions. You can only use each digit once.

a) ┌───────────────┐
   │ 1 2 2 4 6 │
   └───────────────┘

$$\frac{\square}{\square} = \frac{\square}{\square\,\square}$$

b) ┌───────────────┐
   │ 0 1 2 4 5 │
   └───────────────┘

$$\frac{\square}{\square} = \frac{\square}{\square\,\square}$$

c) ┌──────────────────┐
   │ 1 2 4 4 6 6 │
   └──────────────────┘

$$\frac{\square}{\square} = \frac{\square\,\square}{\square\,\square}$$

6.3 Comparing and ordering fractions

1 Use equivalence to find which fraction is greater. Circle the greater fraction in each pair. You can use the bars to help you.

a) $\frac{4}{5}$ or $\frac{6}{10}$

b) $\frac{6}{10}$ or $\frac{11}{20}$

c) $\frac{6}{8}$ or $\frac{5}{10}$

d) $\frac{5}{10}$ or $\frac{2}{3}$

2 Write each set of fractions in order from smallest to largest.

a) $\frac{1}{4}$ $\frac{7}{8}$ $\frac{1}{2}$ $\frac{3}{8}$

b) $\frac{7}{10}$ $\frac{2}{5}$ $\frac{3}{4}$ $\frac{4}{8}$

c) $\frac{5}{6}$ $\frac{3}{4}$ $\frac{2}{3}$ $\frac{4}{5}$

3 Circle the fraction that is in the incorrect place.

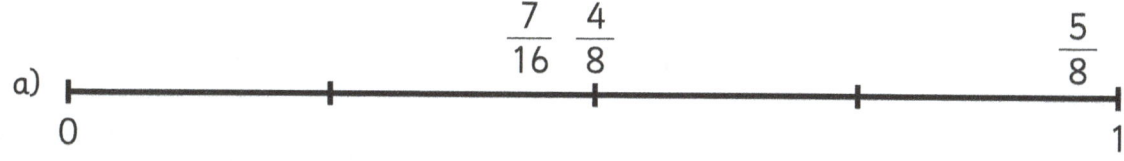

a)

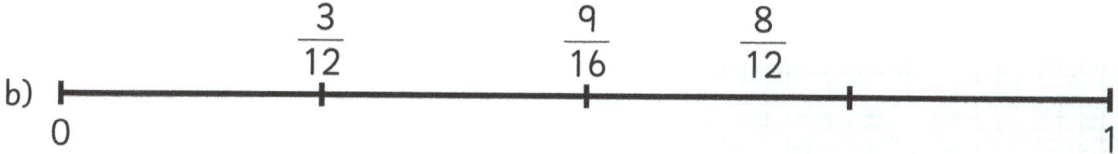

b)

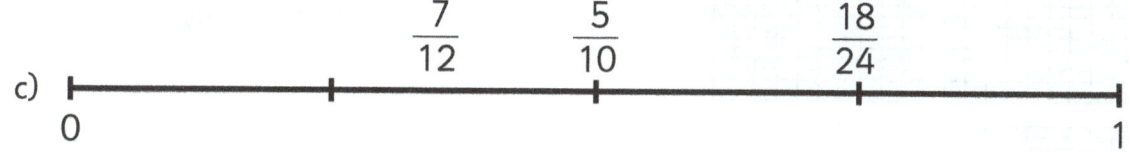

c)

⭐ **Challenge**

a) Tyler gets a large bar of chocolate as a present. He eats one quarter and puts the rest in the cupboard. His mum finds it and eats four sixths of what she finds. How much is left for Tyler?

Draw a picture or diagram to show your thinking.

b) Cullen gets a bar of chocolate too. He eats a third. His dad finds what's left and eats two twelfths. How much is left for Cullen?

Draw a picture or diagram to show your thinking.

1 a) Mark the correct statements with a tick. The first one is done for you.

	Mixed fraction	Hundredths	Decimal fraction
i)	4 wholes and 71 hundredths ✓	471 hundredths ✓	47·1
ii)	3 wholes and 91 tenths	391 hundredths	39.1

b) Write the corrections to part a) below.

i)

ii)

2 Colour the diagram to show the fraction. Write each one as a decimal fraction.

a)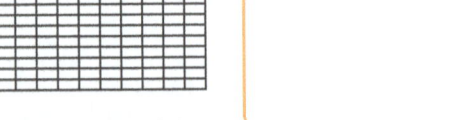

45 hundredths

b)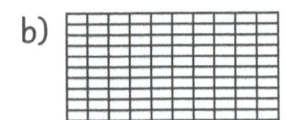

65 hundredths

c)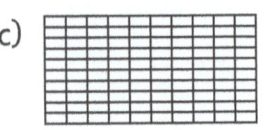

85 hundredths

d)

5 hundredths

3 Write each portion as a decimal fraction. The first one is done for you.

	tenths	hundredths	=	decimal fraction
a)	4	9		0·49
b)				
c)				
d)				

★ Challenge

Each child has £1 in pennies.

Amman	Sandy	Maryam	Freya
$\frac{3}{6}$	$\frac{2}{5}$	$\frac{6}{8}$	$\frac{4}{6}$

Tick the children who can take exactly their fraction of £1 in pennies?

Explain your thinking here and show how many pennies each child gets.

6.5 Decimal equivalents to simple fractions

1 Colour the bars to show tenths. Record the equivalent decimal fraction in the answer box. One has been done for you.

a)

three fifths | six tenths | = 0·6

b)

four eighths | _____

c)

two fifths | _____

d)

six twentieths | _____

2 Use the boxes to help you change each fraction into a decimal fraction.

Write each one as a decimal fraction. The first is done for you.

a) $\dfrac{3}{5}$ = 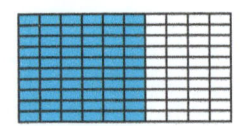 = $\dfrac{6}{10}$ = = 0·6

b) $\dfrac{3}{4}$ = $\dfrac{}{100}$ = =

c) $\dfrac{1}{4} = \dfrac{\boxed{}}{100} =$ [grid] $= \boxed{}$

d) $\dfrac{50}{100} =$ [grid] $= \dfrac{\boxed{}}{100} =$ [grid] $= \boxed{}$

3 Match each fraction to the equivalent decimal fraction. Use the bars to help.

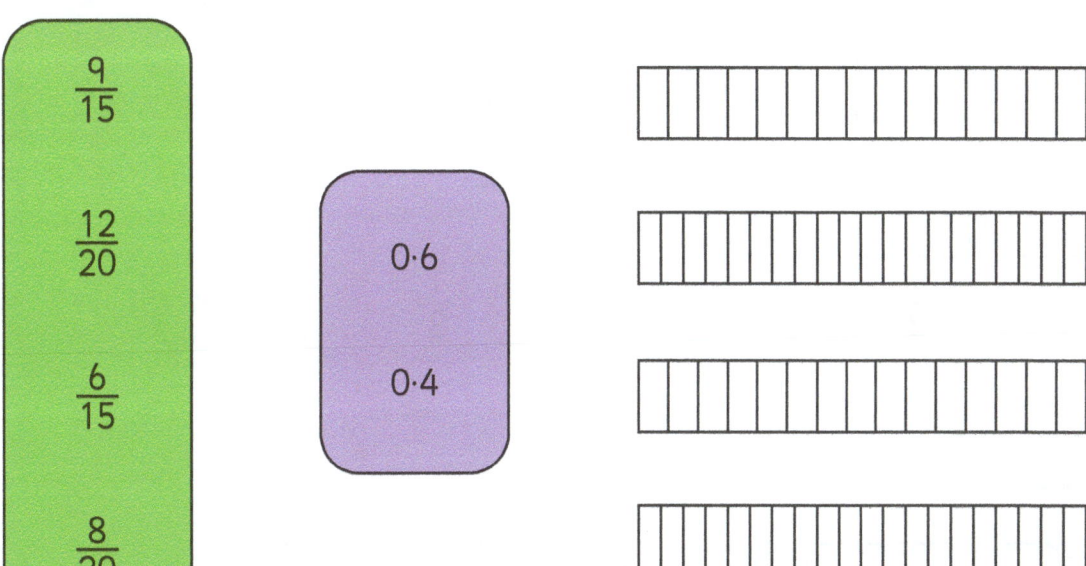

$\dfrac{9}{15}$

$\dfrac{12}{20}$

$\dfrac{6}{15}$

$\dfrac{8}{20}$

0·6

0·4

★ Challenge

Convert these fractions to decimal fractions. Which fraction is the odd one out and why? You may wish to draw or make jottings.

$$\dfrac{16}{20} \qquad \dfrac{40}{50} \qquad \dfrac{15}{30} \qquad \dfrac{20}{25}$$

6.6 Adding and subtracting fractions

1 Solve the following:

a) 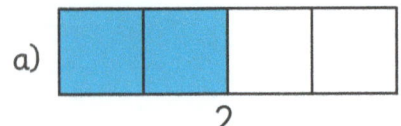 +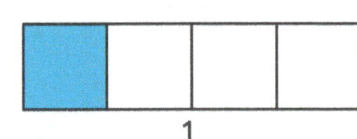

$\frac{2}{4}$ + $\frac{1}{4}$ = ☐

b)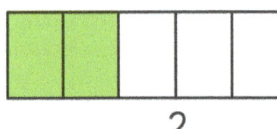

$\frac{5}{6}$ − $\frac{2}{6}$ = ☐

c) 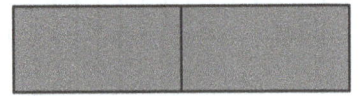 +

$\frac{4}{3}$ + $\frac{1}{3}$ = ☐

d)

$\frac{5}{2}$ − $\frac{2}{2}$ = ☐

2 Solve the following. Use the bar models to help you.

a) $\frac{3}{12} + \frac{4}{12}$ = ☐

b) $\frac{7}{12} - \frac{4}{12}$ = ☐

c) $\frac{3}{12} + \frac{8}{12} =$ [ ]

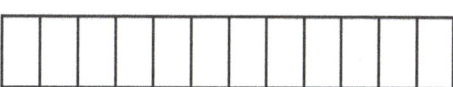

d) $\frac{12}{12} - \frac{3}{12} =$ [ ]

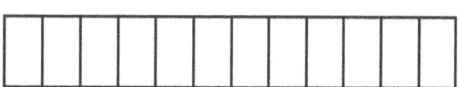

e) $\frac{15}{12} - \frac{12}{12} =$ [ ]

3 Answer the questions. You may want to use the bar models to help you.

a) Jake cuts his apple into sixths and eats four sixths. What fraction of his apple will he have left?

[   ] [ ]

b) Rowan and Blossom are given a bar of chocolate each. They both eat $\frac{3}{6}$ of their chocolate. How much chocolate is eaten altogether?

[   ] [ ]

c) Amir's mum cuts his birthday cake into tenths. Amir eats $\frac{2}{10}$ and his sister eats $\frac{1}{5}$. How much cake is left?

[   ] [ ]

★ Challenge

Use equivalence to solve these. You may want to draw bar models to help you:

a) $\frac{2}{5} + \frac{2}{10} =$ [   ]

b) $\frac{6}{8} - \frac{3}{4} =$ [   ]

1 Draw a line to match the bar model to the correct calculation then calculate the answer.

a) $\frac{3}{10}$ of 1320 = [   ]

| 440 | 440 | 440 |

b) $\frac{3}{5}$ of 1320 = [   ]

| 220 | 220 | 220 | 220 | 220 | 220 |

c) $\frac{3}{6}$ of 1320 = [   ]

| 264 | 264 | 264 | 264 | 264 |

d) $\frac{2}{3}$ of 1320 = [   ]

| 132 | 132 | 132 | 132 | 132 | 132 | 132 | 132 | 132 | 132 |

2 Use the bar models to work out the following.

a)

| 40 | 40 | 40 | 40 |

$\frac{3}{4}$ of 160 = [   ]

b)

(empty bar model with 8 sections)

$\frac{6}{8}$ of 160 = [   ]

c)

(empty bar model with 8 sections)

$\frac{4}{8}$ of 1600 = [   ]

3 Draw a bar model to solve each problem.

a) Mr Cook bakes 736 cupcakes. He gives $\frac{5}{8}$ to his local bake sale. How many does he have left?

b) Ms Shepherd looks after 2624 sheep. She sold $\frac{1}{4}$ of them at market. How many did she sell?

c) Mrs Fisher needs 2500 grams of fish to make a large fish pie. The fishmonger has $\frac{3}{5}$ of the amount she needs. How many grams of fish does she still need?

★ **Challenge**

Four sports clubs share the following information about their members:

Club A has 1240 female members.

Club B has 1116 female members.

Club C has 584 female members.

Club D has 1065 female members.

Use the bar models below to calculate how many members there are in **total** at each club.

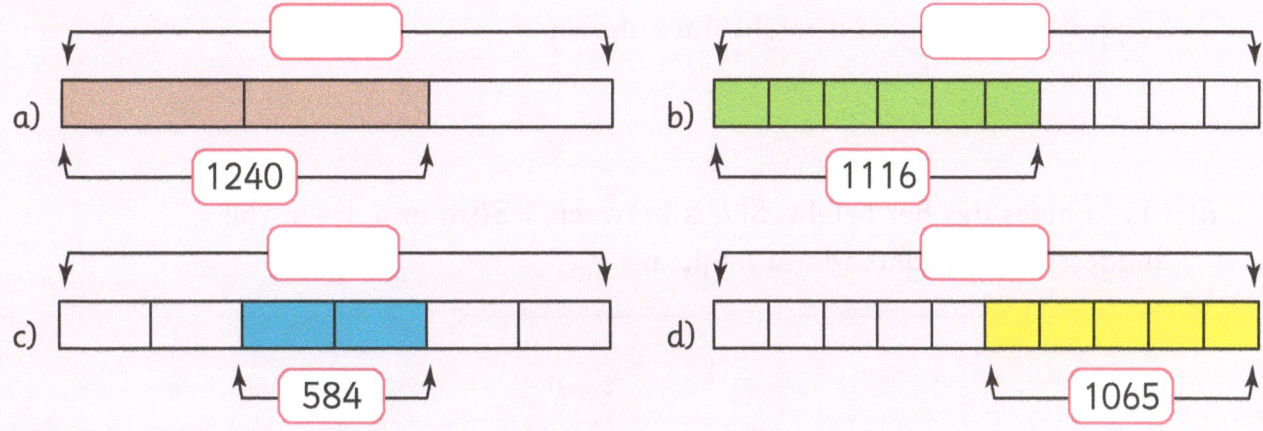

1 Write the correct symbol **<, >** or **=** to make these statements true.

a) 0·5 [ ] 0·45   b) 0·5 [ ] 0·50   c) 1·5 [ ] 1·05

d) 2·5 [ ] 1·55   e) 2·1 [ ] 2·55   f) 2·14 [ ] 21·1

2 Put each set of decimal numbers in order from smallest to largest.

a) 4·14   4·4   4·04   [ ] [ ] [ ]

b) 14·34   14·14   14·4   [ ] [ ] [ ]

c) 414·03   414·3   414·33   [ ] [ ] [ ]

3 a) Jeevan is saving for a new cricket bat. It costs between £79·50 and £80. Suggest three amounts it might cost.

[ ] [ ] [ ]

b) Siobhan gets a personal best for her 25 m swimming race. Her speed is between 28·45 seconds and 28·88 seconds. Suggest three times she might have recorded.

[ ] [ ] [ ]

c) Ryan weighs himself. The scale shows between 32·4 kg and 32·45 kg. Suggest three weights it might have shown.

[ ] [ ] [ ]

d) Maria measures her height. She is between 1·36 m and 1·4 m tall. Suggest three heights Maria might be.

[ ] [ ] [ ]

Complete the grid using these numbers.

2·05   3̶·̶7̶5̶   3·99   0·48   1̶·̶3̶5̶   0̶·̶1̶4̶

3·7   2·3   4̶·̶0̶2̶   1·3   0·36   4·6

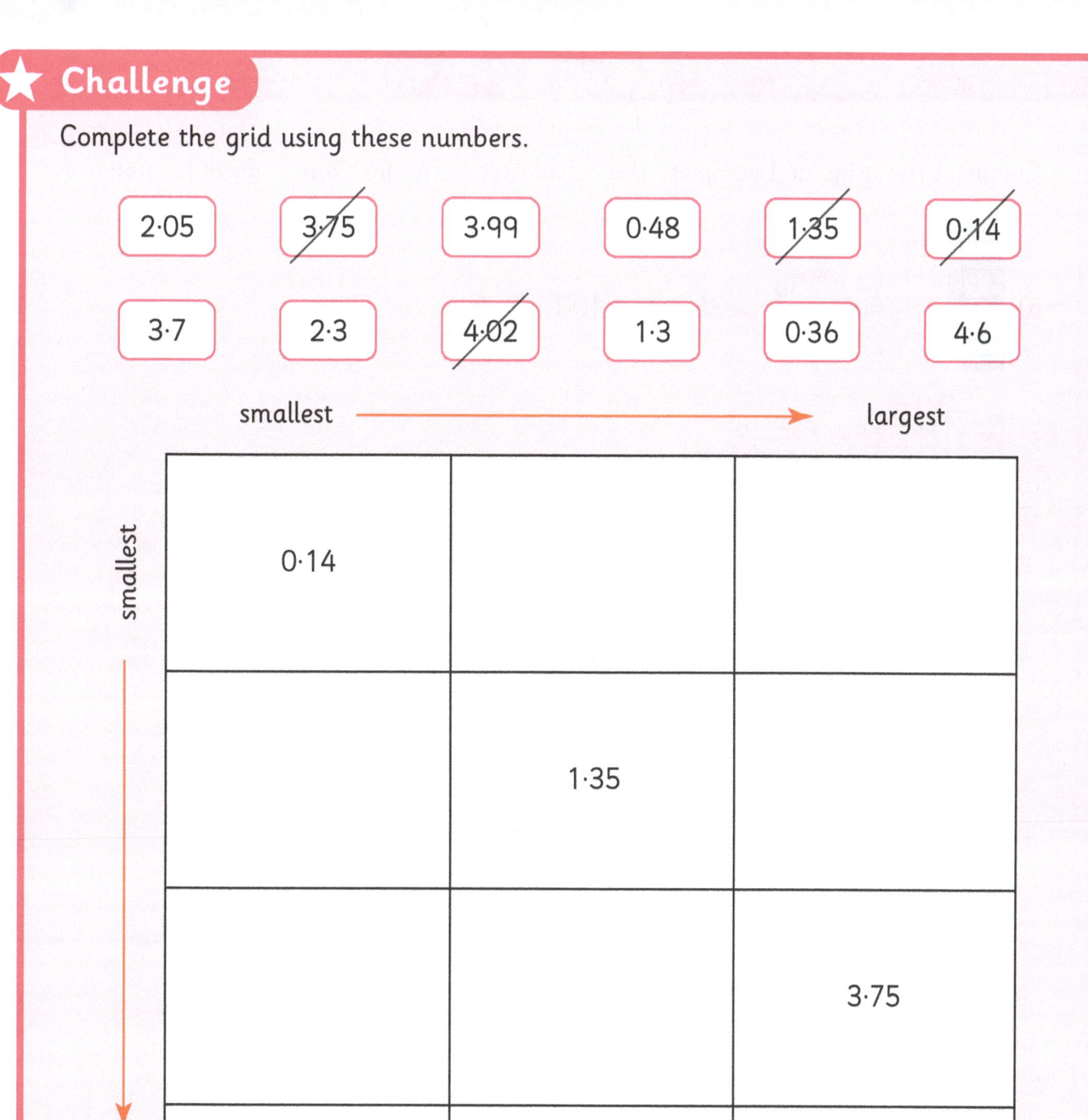

smallest ⟶ largest

smallest

largest

0·14		
	1·35	
		3·75
	4·02	

1 Colour in the grids and complete the calculations. The first one is done for you.

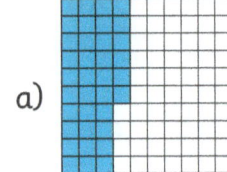

a) $= \dfrac{36}{100} = 36$ out of $100 = 36\%$

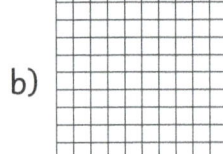

b) $= \dfrac{46}{100} = \boxed{}$ out of $100 = \boxed{}\ \%$

c) $= \dfrac{\boxed{}}{100} = 56$ out of $100 = \boxed{}\ \%$

d) $= \dfrac{\boxed{}}{100} = \boxed{}$ out of $\boxed{} = 66\%$

2 Colour the 100 blocks to show the following:

a) $\dfrac{45}{100}$ blue, 55% red

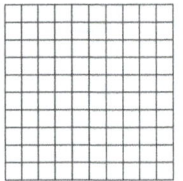

b) $\dfrac{35}{100}$ blue, 65% red

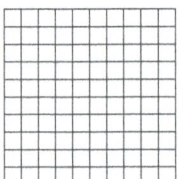

c) $\dfrac{65}{100}$ blue, 35% red

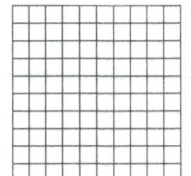

d) $\dfrac{75}{100}$ blue, 25% red

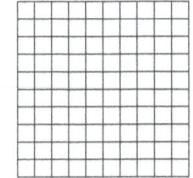

3 Complete the following calculations. The first one has been done for you.

a) $5\% = 5$ out of $100 = \dfrac{5}{100}$

b) $10\% = $ ⬚ out of $100 = $ ⬚

c) $15\% = $ ⬚ out of $100 = $ ⬚

d) $20\% = $ ⬚ out of $100 = $ ⬚

e) $25\% = $ ⬚ out of $100 = $ ⬚

⭐ **Challenge**

a) Sort the fractions into those that can have an equivalence with hundredths and those that cannot. Write your answers in the table below. You can use the grids to help you.

$\dfrac{8}{20}$ $\dfrac{9}{10}$ $\dfrac{8}{50}$ $\dfrac{8}{40}$ $\dfrac{9}{60}$ $\dfrac{8}{30}$ $\dfrac{9}{40}$

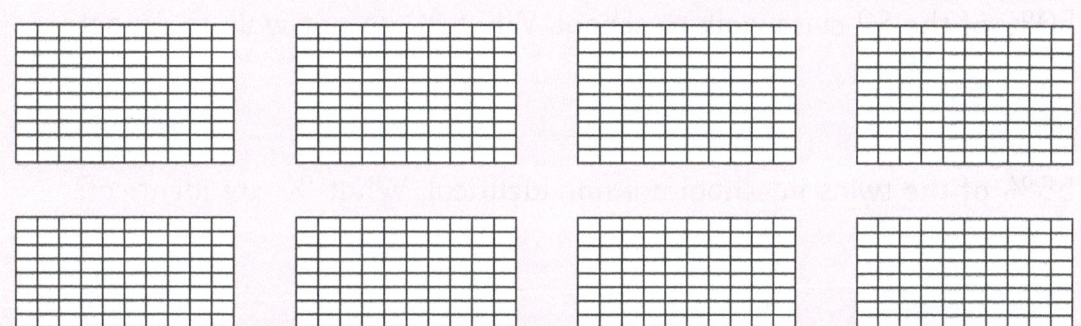

Has an equivalence with hundredths	Does not have an equivalence with hundredths

b) Add a fraction of your own to each side of the table.

6.10 Converting fractions to percentages

1 Colour the grids and complete the fractions:

a) $\dfrac{1}{2}$ = 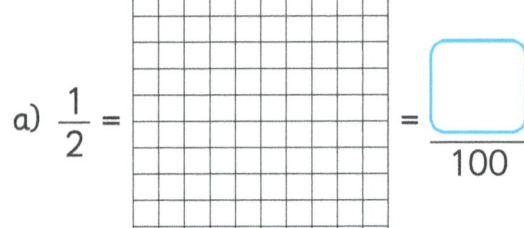 = $\dfrac{\boxed{}}{100}$

b) $\dfrac{1}{4}$ = 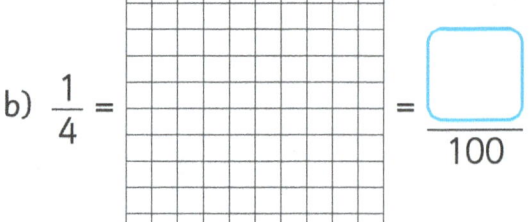 = $\dfrac{\boxed{}}{100}$

c) $\dfrac{1}{5}$ = 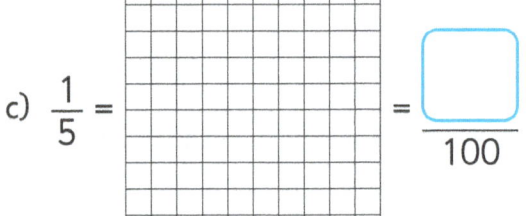 = $\dfrac{\boxed{}}{100}$

d) $\dfrac{2}{5}$ = 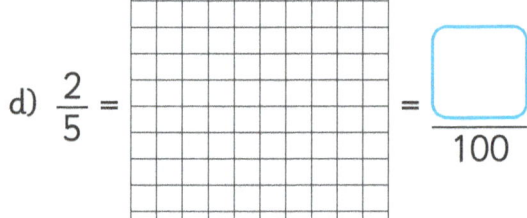 = $\dfrac{\boxed{}}{100}$

2 The students have been looking at student data in their school.

a) 50% of the S2 class walk to school. What % do not walk to school?

b) 55% of the twins in school are non-identical. What % are identical?

c) 5% of the pupils in S3 are left-handed. What % are right-handed?

d) 25% of S5 pupils speak more than one language. What % speak only one language?

3 Use these numbers to create six different fractions and convert them to percentages. One has been done for you.

Numerator $\boxed{4}$ or $\boxed{8}$

Denominator $\boxed{10}$ or $\boxed{20}$ or $\boxed{50}$

a) $\dfrac{\boxed{4}}{\boxed{10}} = \dfrac{\boxed{40}}{100} = \boxed{40}$ %

b) $\dfrac{\boxed{}}{\boxed{}} = \dfrac{\boxed{}}{100} = \boxed{}$ %

c) $\dfrac{\boxed{}}{\boxed{}} = \dfrac{\boxed{}}{100} = \boxed{}$ %

d) $\dfrac{\boxed{}}{\boxed{}} = \dfrac{\boxed{}}{100} = \boxed{}$ %

e) $\dfrac{\boxed{}}{\boxed{}} = \dfrac{\boxed{}}{100} = \boxed{}$ %

f) $\dfrac{\boxed{}}{\boxed{}} = \dfrac{\boxed{}}{100} = \boxed{}$ %

★ Challenge

Alice says that you can tell if a fraction can convert to a percentage if you know the denominator. Simone says you need to know the numerator **and** the denominator. Show some examples to prove who is correct and why.

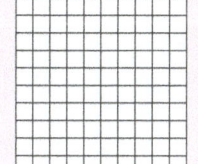 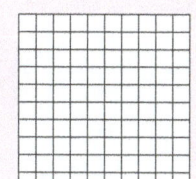 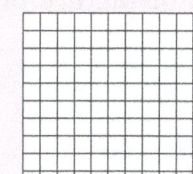 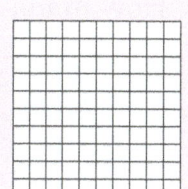

6.11 Percentage calculation

1 Use the bar models to calculate the percentage of students for each school who walk to school or travel by car.

Glenfield School – 100 students

a)

walk

car

Seaview School – 90 students

b)

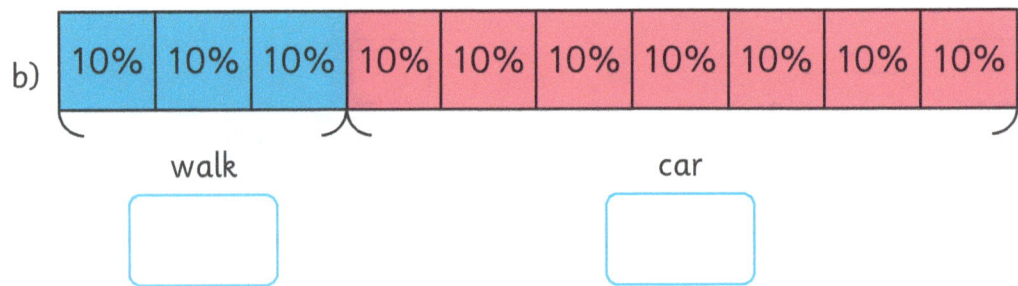

walk

car

2 Draw a bar model to solve each of these.

a) Sami hands a bag of 10 books into the charity shop. 60% of them are picture books. How many are not picture books?

b) Lilya's parents invite 80 people to a party. 20% reply to say they cannot make it. How many guests can Lilya's parents expect?

3 Fieldview School has 120 students.

a) At lunch time, 10% of students go home. How many students go home?

b) At lunch time, 30% of students have a vegetarian school dinner. How many students have a vegetarian dinner?

c) At lunch time, 40% of students have a non-vegetarian school dinner. How many students have a non-vegetarian dinner?

d) At lunch time, the rest of the students have a packed lunch. How many students have a packed lunch?

What percentage of the school is this?

⭐ Challenge

Some students are having a netball competition. Who has the most success? Explain your answer.

Pat tries 30 times and gets 15 balls in the net.

Helena tries 25 times and gets 12 balls in the net.

Leo tries 40 times and gets 24 balls in the net.

Sheena tries 20 times and gets 14 balls in the net.

7.1 Money problems using the four operations

1 Complete the table. The first row is done for you.

Amount	Pounds and pence	Pence
Three pounds sixty	£3·60	360p
Three pounds six		
	£4·08	
		480p
Fourteen pounds ten		
	£24·20	

2 Answer the following questions. You may prefer to convert the amounts to pence to work out the calculations. Give your answers in pounds and pence.

a) i) The hairdresser charges £22 for Hannah to have her hair cut. Hannah also buys a bottle of shampoo for £3·45 and pays the hairdresser. How much does she pay in total?

ii) Hannah gets £14·55 in change. How much did she give the hairdresser?

iii) What notes might she have paid with? Show three ways.

b) Hannah and her mum go to a café after the hair cut. They both have a milkshake and buy a fudge brownie to share.

<table>
<tr><td colspan="2">Menu</td></tr>
<tr><td>Milkshake</td><td>£2</td></tr>
<tr><td>Fudge Brownie</td><td>£4·50</td></tr>
<tr><td>Scone</td><td>£1·79</td></tr>
</table>

Mum pays with a note and three coins. What note and coins did she use?

⭐ **Challenge**

Four students do a sponsored 10 km walk.

- Marc gets £2·25 for each kilometre.
- Lisa gets £3·05 for each kilometre.
- Fiona gets £4·35 for each kilometre.
- Conor gets £2·45 for each kilometre.

a) How much does each student make?

b) How much do the students make altogether?

1 Sarah is shopping for dog treats. She wants to get the best deal. Circle the best deal for each treat and explain your thinking.

Item	Molly's Market	Patrick's Pets	Reason
a) Chew Bones	10 for £3·99	20 for £7·50	
b) Doggie Drops	Pack of 100 for £6·50	Pack of 10 for 70p	
c) Pet Dental Sticks	Pack of 20 for £1·99	Pack of 40 for £3·00	

2 Solve the following:

a) A cinema ticket is £6·00. A medium bag of popcorn is £5·00. A monthly ticket is £25 and includes one free bag of popcorn and free entry to the cinema four times. Emma knows she wants to go to three films in January and she likes a bag of popcorn each time. Should she get the monthly ticket or not? Explain your thinking.

b) Cornflakes come in two sizes at the supermarket: a 1 kg box for £1·20 or a 500 g box for 70p. How much will Sia save if she buys the best deal?

c) Dad is buying cupcakes for Sam's party. He needs 30 cupcakes. A local baker will bake them for £1·12 each. She will put Sam's name on each and decorate them in Sam's favourite colour. The supermarket has packs of six cupcakes for £6·60. Who should Dad buy the cupcakes from and why?

★ Challenge

Some of Harry's family are going on holiday. There is one adult and three children.

Zippy Holidays

Return flights: £180 per adult
£120 per child

Hotel: £600 for family room

Breakfast: £160 for the stay

Flying Fox Holidays

Return flights: £500 for 4 people

Hotel: £850 for family room (includes Breakfast)

a) Which deal offers the best value for Harry's family?

b) If the family buys the best deal and saves £100 each month, how long will it take them to have enough money to pay for the holiday?

1 Decide whether each sale gives a profit or a loss.

Item	Bought	Sold	Profit or loss?
Car	£28 490	£18 000	
House	£399 405	£410 245	
Mobile phone	£449·00	£350·00	
Concert ticket	£105·00	£120·00	

2 The school tuck shop buys a box of 32 packets of crisps for £19. The staff sell the packets of crisps for 70p each.

a) Does the school make a profit or a loss?

b) How much did they make or lose?

3 Mrs Logan buys a new coat for £77 in January. In July she sells it for £55.

a) Did she make a profit or a loss?

b) How much did she make or lose?

4 Mr Good and Mr Sharp buy a flat. They pay £62000. They spend £2400 on repairs. They sell the flat a year later for £82000.

a) Did they make a profit or a loss?

b) How much did they make or lose?

★ Challenge

Marta and Isabella are raising money for charity and are planning a prom. They have set themselves a target of raising £400.

- They get some decorations for free.
- They pay £50 to hire the hall from 7 pm–11 pm.
- The prom costs £275 for 4 hours.

a) How much do they need to raise to make a profit?

b) How much do they need to raise to meet their charity target?

c) If 100 students buy a ticket, how much must the ticket cost so that Marta and Isabella make their target?

7.4 Discounts

1 Work out the new price for each of these items.

Item	Cost	Discount	New cost
	£3·80	50% off	
	£1·20	Half price	
	£2·00	10% off	
	£1·60	25% off	

2

£8·00   £6·00   £20·00   £10·00

a) Mr Bigg has a sale in his shop. He puts a big sign in the window: 10% off all goods.

How much does each item cost now?

b) Mrs Bigg thinks the sale should be better. She takes down the 10% off sign and puts a 25% off sign up.

How much does each item cost now?

3 In the sale, David pays £3·05 for a cap which is half price. How much did the cap cost originally?

⭐ **Challenge**

a) There is a sale at Bob's Bargains. Lacey wants a video game which has 25% off and is now £18. What was the original price? Show your working.

b) Andrew wants a board game which has 75% off and is now £9. What was the original price? Show your working.

7.5 Credit, debit and debt

1 Look at Mrs Smith's bank statement and complete the balance column.

	(£) In	(£) Out	(£) Balance
			3500
Café		22·00	
Petrol		88·00	
Gas		216·00	
Shopping		125·00	
Pay	600·00		
Council tax		350·00	

Is she in debt or in credit at the end?

2 Look at Mr Smith's bank statement and complete the balance column.

	(£) In	(£) Out	(£) Balance
			500
Petrol		40·00	
Birthday	50·00		
Shopping		122·00	
Energy		215·00	
Clothes		60·00	
Car		152·00	
Dog food		27·00	

Is he in debt or in credit at the end?

3 Look at the information for Mr Ahmed's credit card account.

Credit limit £2000

Flights	£600·00
Concert tickets	£120·00
PAYMENT – THANK YOU	£900·00 CR
Train	£125·00
Restaurant	£97·00
Petrol station	£77·00
Cinema	£34·00
Present balance	
Available to spend	

Calculate Mr Ahmed's balance and how much he has to spend. CR means that a payment has been made.

★ Challenge

You can borrow £400 with your credit card. Think of something you would like to buy.

What is the item and the cost?

a) If you do not pay your credit card in time, you will have to pay an extra 10% of the amount you owe. How much is this?

b) Explain a different way to pay for your item.

1 Complete these:

a) $1 + 12 = \boxed{}$

b) $11 + 12 = \boxed{}$

c) $\boxed{} + 12 = 22$

d) $\boxed{} + 12 = 21$

e) $8 + \boxed{} = 20$

f) $7 + \boxed{} = 19$

g) $\boxed{} + 12 = 17$

h) $\boxed{} + 12 = 18$

2 Fill in the missing equivalent times:

12-hour clock	24-hour clock
2·00 pm	1400
2·30 pm	
2·45 pm	
3·45 pm	
4·30 pm	
5·30 pm	
6·30 pm	
	1845
	1945
	2000
	2100
	2130
	2230
	2330

3 Write these times in the table below. Start with the earliest time and write them in order.

a) 0801   b) 8·01 pm   c) 2000   d) 8·00 am   e) 12·01 am

f) 0002   g) 12·03 pm   h) 1204   i) 2359   j) 1159

Between midnight and noon	Between noon and midnight

4 Put the following in time order starting with the time in bold:

a) 3·00 pm   1440   1240   **11·40 am**

b) 12·50 am   **0040**   1040   10·40 pm

c) **1600**   4·15 pm   0415   4·00 am

⭐ **Challenge**

Conor is meeting Muhammad at the park. He wrote the time on a piece of paper but he has lost it. He remembers writing down **0** twice and a **4**. He knows it was in the afternoon. He remembers using 24-hour time. What time is Conor supposed to meet Muhammad?

1. This number line and the clock face both show 60 minutes.

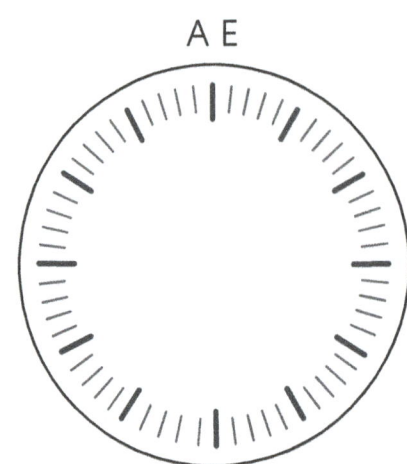

Write the following on the number line and the clock face and answer the questions.

A at zero minutes

C at the halfway point. Half of 60 = ☐

B halfway between **A** and **C**. One quarter of 60 = ☐

D halfway between C and E. Three quarters of 60 = ☐

$\frac{1}{4}$ of an hour = ☐ minutes.

$\frac{1}{2}$ an hour = ☐ minutes.

$\frac{3}{4}$ of an hour = ☐ minutes.

1 hour = ☐ minutes.

2 Answer the following calculations:

a) $0900 + \frac{1}{2}$ an hour = ☐ b) $0915 + \frac{1}{2}$ an hour = ☐

c) $0930 + \frac{1}{2}$ an hour = ☐ d) $7.00\,pm + \frac{3}{4}$ of an hour = ☐

e) $7.15\,pm + \frac{3}{4}$ of an hour = ☐ f) $7.30\,pm + \frac{3}{4}$ of an hour = ☐

3 Draw the correct time on the clock faces.

a) Suzi is going to meet Dee at 2·20 pm. It takes her half an hour to walk to Dee's house. When should she set out?

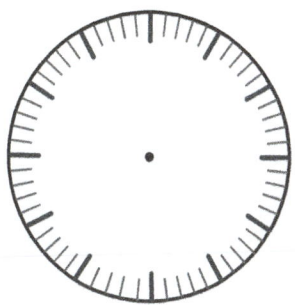

b) Marcus sets off for school at 8·25 am. He takes a quarter of an hour to get to school. When does he arrive?

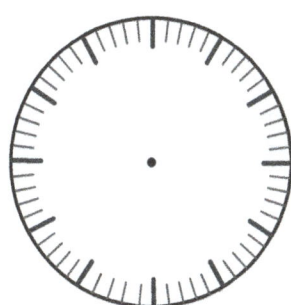

c) At school the students get three quarters of an hour for lunch. If lunch break starts at 12·40 pm, when will it end?

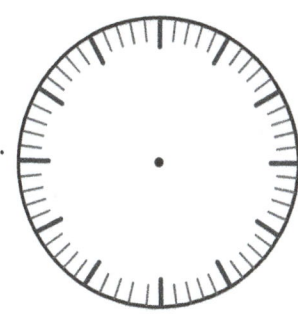

You need to play this game with a partner or play as two different people!
Use a dice.

$\frac{1}{4}$ hour   15 minutes $\frac{1}{2}$ an hour   30 minutes $\frac{3}{4}$ hour   45 minutes

Both players start at 9.00 am. Roll the dice and add the matching amount of time. For example, if you roll a 3 you add $\frac{1}{2}$ an hour. Keep a running total. The winner is the first player to reach 3 pm.

Player 1	Player 2
9.00 am	9.00 am

8.3 Calculating time intervals using timetables

1 Use the timetable to answer the questions.

Channel 1	Channel 2	Channel 3
1610 Mr Moon	**1600** Cannonball	**1700** Evening News
1620 Antman	**1610** Doris	**1725** Weather
1640 Beat the Teacher	**1620** Power Pets	**1730** Snooker
1655 News Time	**1635** Magic Tunes	**1830** Health Hotel
1705 Red Robin	**1650** Space Adventure	**1915** Who Eats What?
1735 Quizzer	**1700** News and Views	**1945** Greenside
1800 News	**1730** Fashion Fun	**2030** Bingo Balls
1835 Scotland Latest	**1750** Cat and Carrot	**2100** New House Old House
1900 Bill and Bob	**1810** Weather Report	**2200** Late News

a) Abigail starts watching TV at 1600 and went to have tea at the end of Red Robin. How long was she watching television for?

b) Petr watches Antman and Space Adventure. How long does he watch TV?

c) Zadie watches a news programme that lasts longer than half an hour. Which channel is she watching?

d) Jayce starts to watch Channel 3 at 1730. He watches TV for 90 minutes. What programme is on TV when he switches off the telly?

e) What programmes could these people be watching?

Name	Channel	Length of programme	Name of programme
Billy	Channel 2	15 minutes	
Alisha	Channel 3	45 minutes	
Cameron	Channel 1	10 minutes	

2 Here is a train timetable for trains from Edinburgh to Glasgow. Calculate the journey time for each train.

	A	B	C	D	E
Edinburgh	1208	1239	1245	1312	1315
Glasgow	1330	1400	1337	1430	1406

Train A: Train B: Train C: Train D: Train E:

⭐ **Challenge**

a) Complete the bus timetable for the Number 72 bus route.

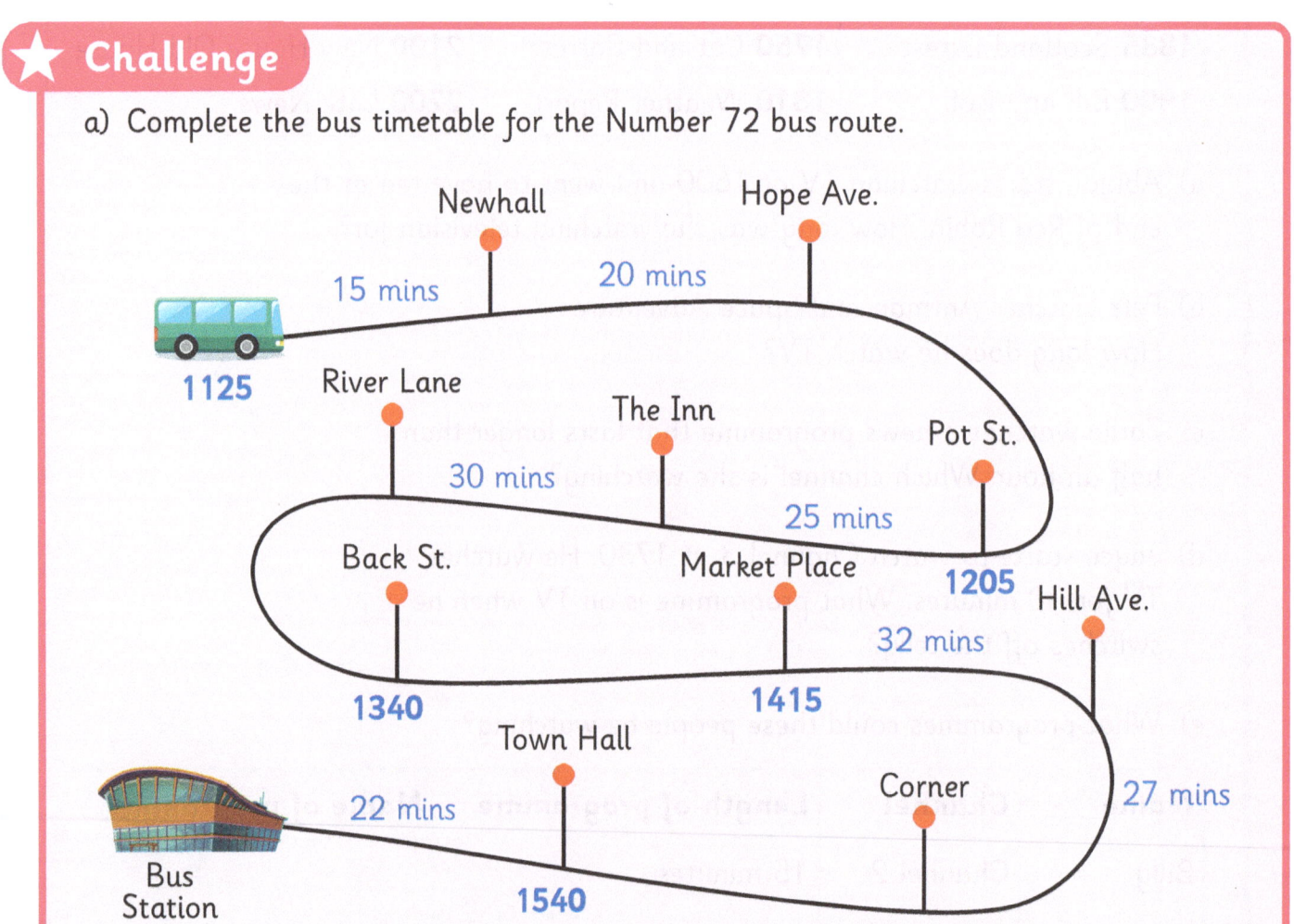

Bus stop	Time	Length of journey (minutes)
Start	1125	
Newhall	1140	15
Hope Ave		
Pot St		
The Inn		
River Lane		
Back St		
Market Place		
Hill Ave		
Corner		
Town Hall		
Bus Station		

b) Draw your own bus route and timetable.

8.4 Measuring time

1 Which unit would be best for measuring time in these situations. Tick the correct box.

	seconds	minutes	hours	days
Saying your name aloud				
The Olympic Games				
A film at the cinema				
A song on the radio				
Getting dressed for school				

2 For each activity estimate how long it will take you to do it, then time yourself. Record your estimate and time in the table.

	Estimate more or less than a minute	Actual time
Writing your name backwards		
Walking to the door and back		
Reading a page of a book		
Putting your jacket on		
Asking five friends their favourite colour		

3 Make a list of all the devices in your house and/or school that can be used to measure time.

Read the list below. You are going to create a birthday card for a friend.
Time how long it takes to do each part.

Activity: Create a birthday card	Time
Getting resources together	
Folding card	
Drawing picture	
Colouring picture	
Writing message	

HAPPY BIRTHDAY

8.5 Speed, distance and time calculations

Can you solve these problems using what you know about the link between speed, distance and time?

1. Mr Thomson drives at 20 miles per hour (mph) for 30 minutes. How far does he travel?

2. Ms Khan drives at 60 mph for $2\frac{1}{2}$ hours. How far did she travel?

3. Mrs Lee runs for $1\frac{1}{2}$ hours at a speed of 10 mph. How far did she run?

4. Mr Choi drives at 50 mph and travels 50 miles. How long was he driving for?

5. Mrs Gray drives at 20 mph and travels 5 miles. How long was she driving for?

6. Ms Anderson cycles at 15 mph and travels 30 miles. How long was she cycling for?

a) Measure a distance of 10 metres and mark it. You will need a calculator. Use this to work out your speed for these:

To find the speed travelled in kilometres per hour, follow these steps:
1. Ensure your estimated time to travel 1000 m is in minutes.
2. Divide 60 (minutes) by your estimated time to travel 1000 m.

	Time to travel 10 metres	Time to travel 100 metres	Estimated time to travel 1000 m (1 km)	Speed travelled in kilometres per hour
Hopping				
Running forwards				
Walking backwards				
Jumping				

b) Measure a distance of 20 metres and mark it. Use this to work out your speed for these:

Activity	Time to travel 20 metres	Time to travel 100 metres	Estimated time to travel 1000 m	Speed travelled in kilometres per hour
Hopping				
Running forwards				
Walking backwards				
Jumping				

8.6 Time problems

1 Nick walks 25 minutes to Lizzi's house and waits 15 minutes for her to get ready. They then walk 25 minutes to meet Johnny. How long was it until Nick and Johnny met up? Write your answer in hours and minutes.

2 Four students run the book stall at the fair. The fair lasts from 10 am until 12·40 pm How long does each student have to look after the stall if they all do the same amount of time?

3 Sabrina was 12 minutes late for her dance class. The class lasts an hour and a quarter. How long did Sabrina spend at that class? Write your answer in hours and minutes.

4 Brona wants to swim 200 metres in under 6 minutes. Her latest time is 340 seconds. Has she been successful? Explain your answer.

5 Roisin and Ana go shopping on Saturday. They leave at 1430. The bus journey into town takes 10 minutes and they spend 90 minutes shopping. They go to a café for 35 minutes and then return home. What time do they return home?

6 Jakob walks to and from school twice a week. The journey one way takes him 25 minutes. How long does Jakob take walking to and from school in a fortnight?

★ **Challenge**

Write the date and time now. Date: [   ]   Time: [   ]

Use the information above to answer the following:

a) In ten minutes the time will be [   ]

b) In twenty minutes the time will be [   ]

c) In 240 seconds the time will be [   ]

d) In 24 hours the date and time will be [   ]

e) In 8 days the date will be [   ]

f) In 2 months the date will be [   ]

g) In 48 days the date will be [   ]

1 Use a ruler to draw these lines. Then write their lengths in centimetres.

a) 60 mm

[ ] cm

b) 69 mm

[ ] cm

2 Fill in the distances in kilometres. Use the most direct route each time.

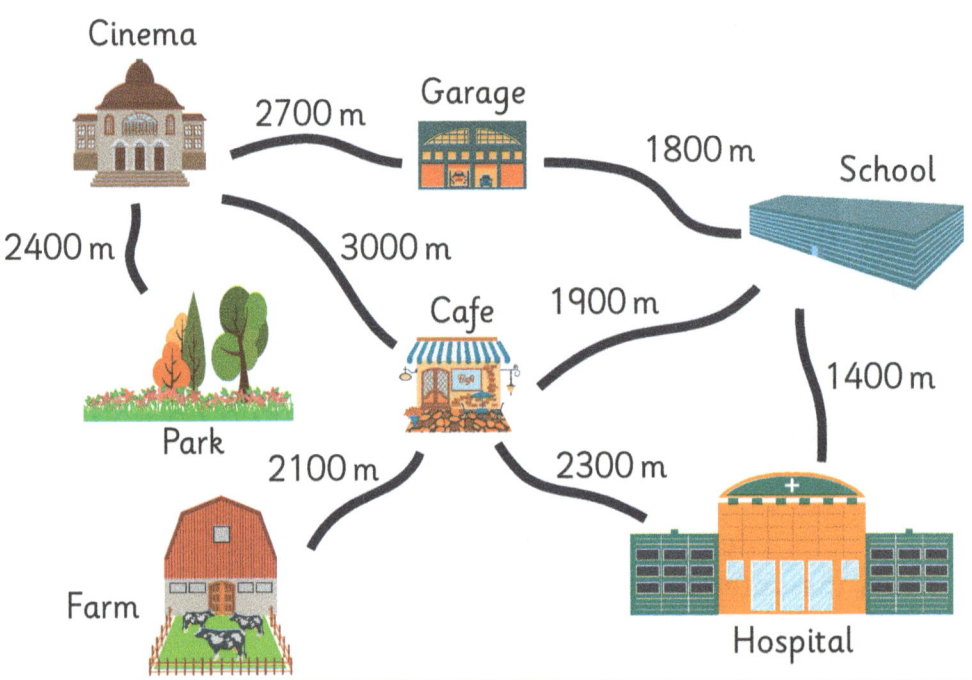

From	To	Distance in km
hospital	school	
farm	cafe	
cinema	park	
garage	cinema	
garage	school	
cinema	hospital	
farm	school	
cafe	park	

3 You will need a measuring tape, something to throw and your workbook. Go to a place you can throw your object. Estimate and measure a distance of 1·4 m. Throw your object and measure where it lands.

Object	Distance where it lands in cm
First throw	
Second throw	
Third throw	
Fourth throw	

★ Challenge

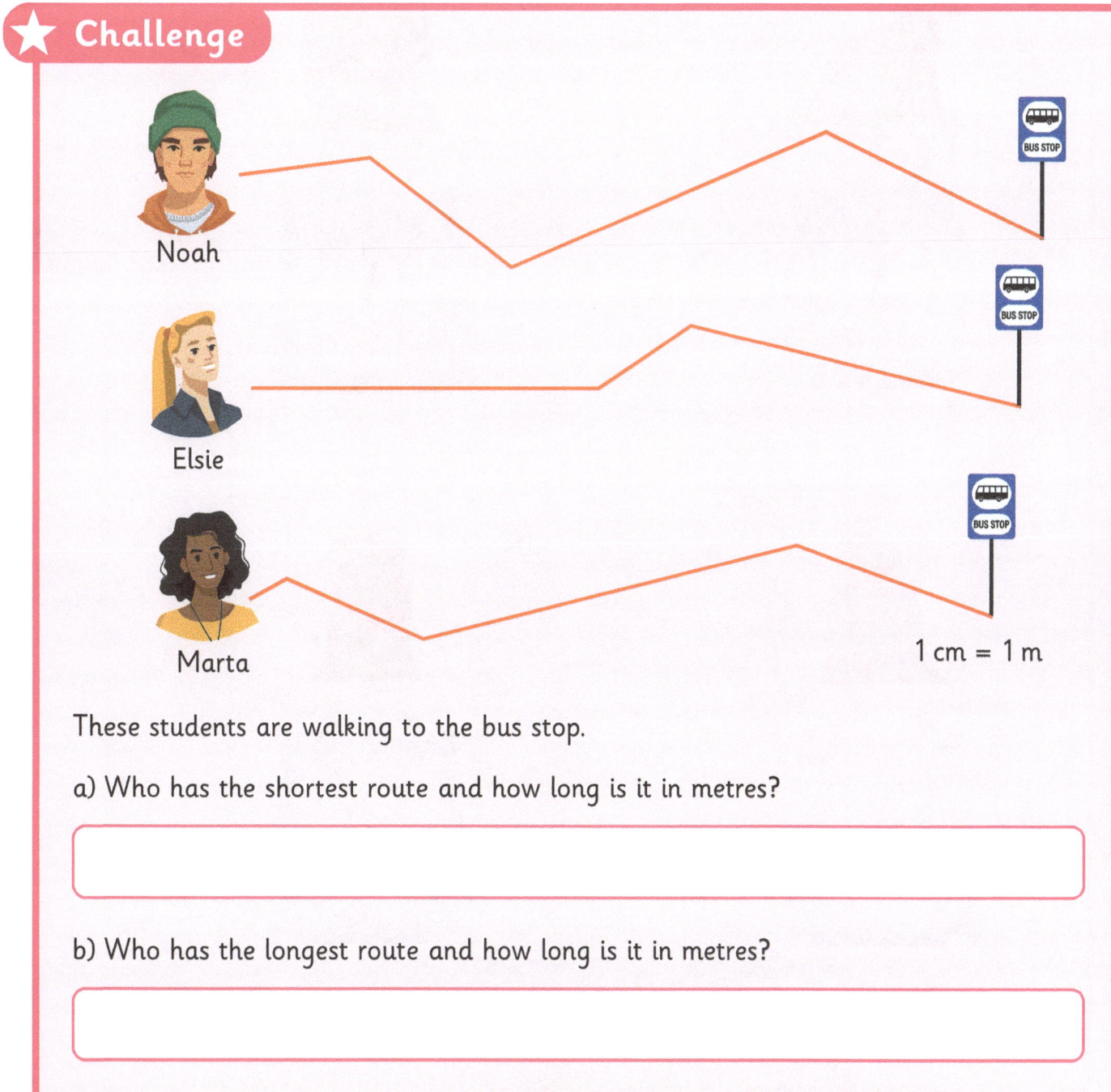

These students are walking to the bus stop.

a) Who has the shortest route and how long is it in metres?

b) Who has the longest route and how long is it in metres?

9.2 Estimating and measuring mass

1 A kilogram can be divided into ten equal parts. Complete the diagram.

100 g			400 g						
0·1 kg						0·7 kg			

2 Write the mass of each sack in grams and kilograms.

a)

b)

c)

d)

3 a) Find three items that you estimate to have a mass of more than 1 kg but less than 1·5 kg.

Item	Estimated mass in kg	Actual mass in kg

b) Put the items in order from lightest to heaviest.

⭐ **Challenge**

a) Use cardboard, paper and tape to build a bridge. It must hold 0·25 kg. Draw a diagram of your bridge.

b) Keep adding weights until your bridge collapses. How much did it manage to hold **before** it collapsed?

9.3 Estimating and measuring capacity

1 Write the amount of liquid in each jug.

a)
ml
5000
4500
4000
3500
3000
2500
2000
1500
1000
500
0

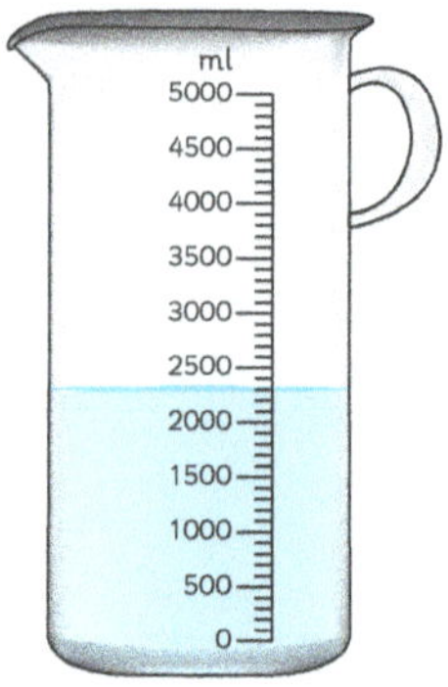

[   ] millilitres

[   ] litres

b)
ml
5000
4500
4000
3500
3000
2500
2000
1500
1000
500
0

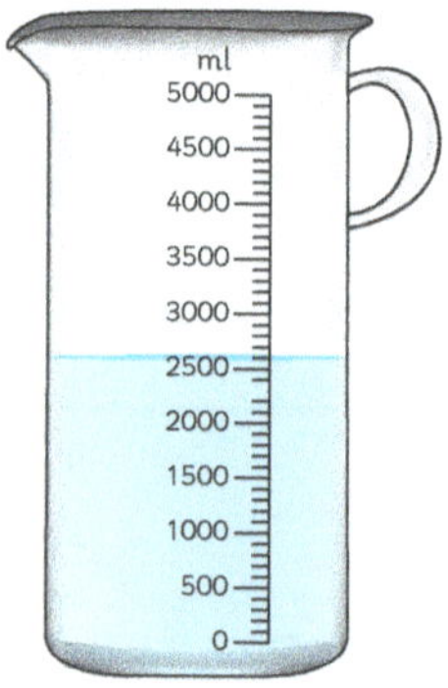

[   ] millilitres

[   ] litres

c)
ml
5000
4500
4000
3500
3000
2500
2000
1500
1000
500
0

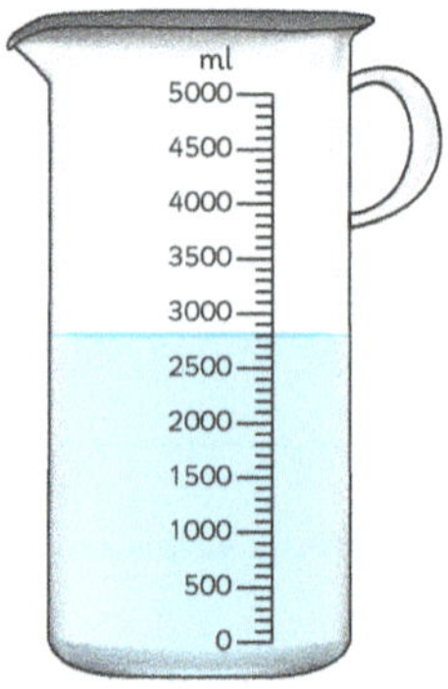

[   ] millilitres

[   ] litres

d)
ml
5000
4500
4000
3500
3000
2500
2000
1500
1000
500
0

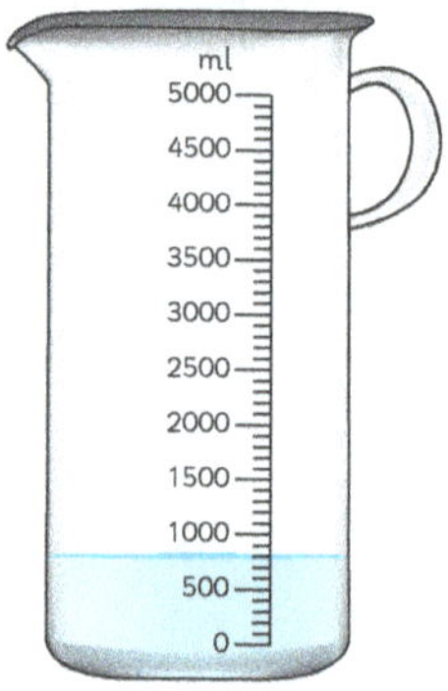

[   ] millilitres

[   ] litres

2 Fill with the correct amount of liquid by colouring these jugs.

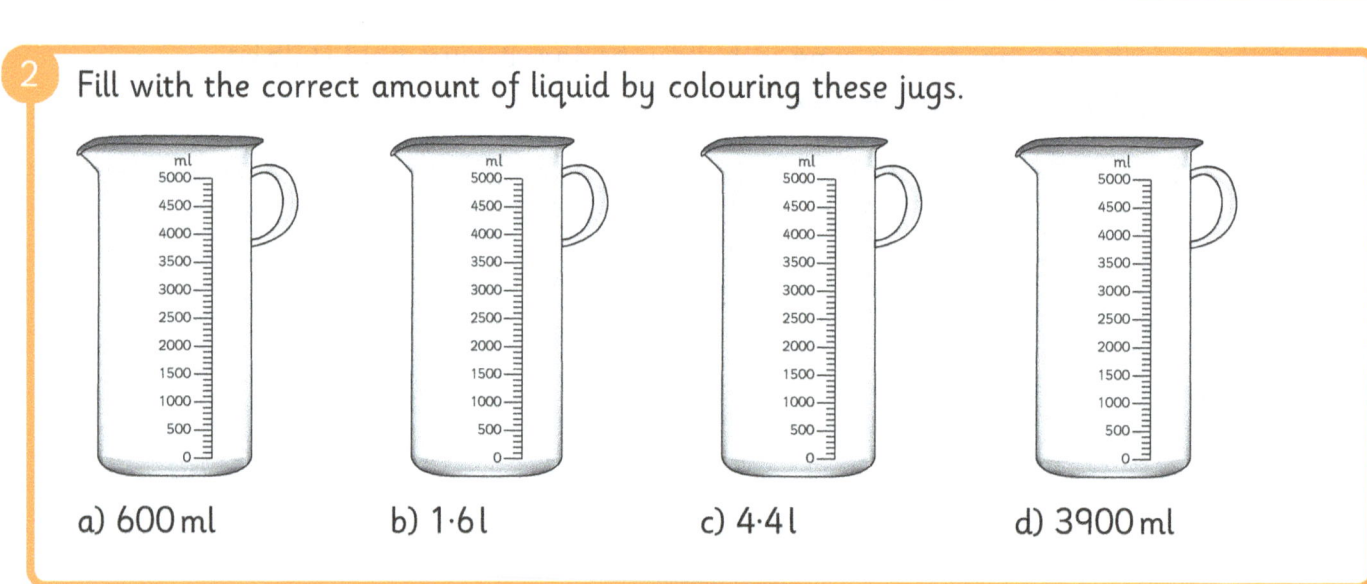

a) 600 ml   b) 1·6 l   c) 4·4 l   d) 3900 ml

3 Draw lines to match these measurements.

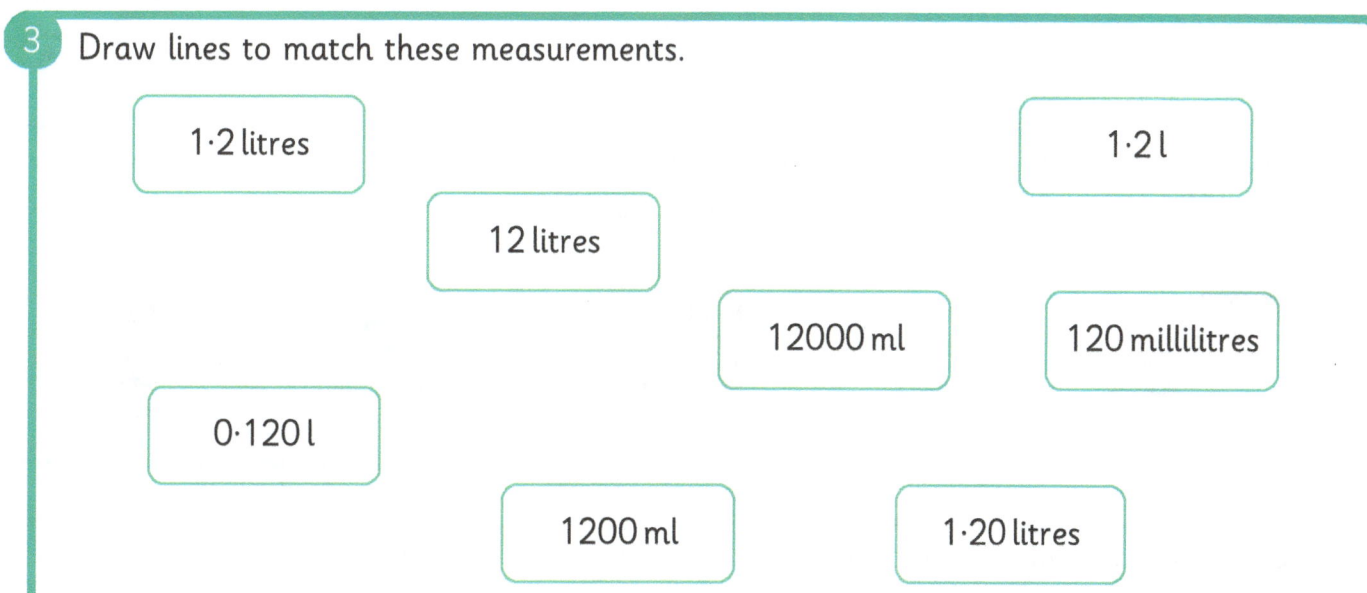

1·2 litres

1·2 l

12 litres

12000 ml

120 millilitres

0·120 l

1200 ml

1·20 litres

★ **Challenge**

Joan has an empty bottle that holds 4 litres of water. Dylan has an empty bottle that holds 7 litres of water. They have access to a sink. Rob challenges them to bring him 5 litres using only their two bottles.

How can Joan and Dylan get 5 litres using only their bottles?

4 litres

7 litres

1 Capacity measures the amount a container can hold. Volume is the space taken up by a 3D shape. Circle the items below that can have capacity.

2 Complete the table. The first one is done for you.

Cubic centimetres		millilitres
45 cm³	is the same as	45 ml
55 cm³	is the same as	
	is the same as	550 ml
505 cm³	is the same as	
	is the same as	1 litre

3 Imagine each of these filled with cubic centimetres. Write in the volume and capacity.

a)

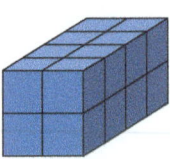

2 cm
4 cm
2 cm

Volume = _____ cm³

Capacity = _____ ml

b)

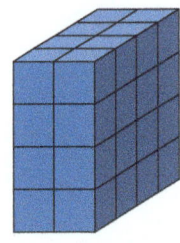

4 cm
4 cm
2 cm

Volume = _____ cm³

Capacity = _____ ml

c)

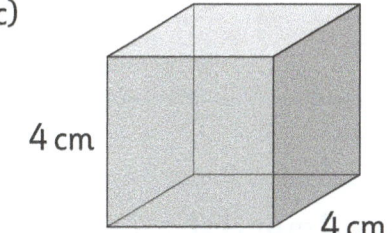 

4 cm
4 cm
4 cm

Volume = _____ cm³

Capacity = _____ ml

d)

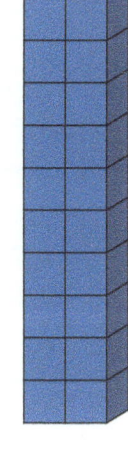

10 cm

2 cm 1 cm

Volume = _____ cm³

Capacity = _____ ml

e)

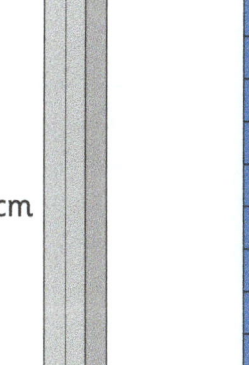

10 cm

1 cm 1 cm

Volume = _____ cm³

Capacity = _____ ml

★ Challenge

Nuria builds a cube that is 10 cm by 10 cm by 10 cm.

She knows that $10 \times 10 \times 10$ is 1000. She also knows there are 100 cm in a m so there are 10 m³ in 1000 cm³ so thinks this means the volume of her cube is 10 m³.

Matthew disagrees!

Who is correct? Explain your reasoning.

9.5 Imperial measurement

1. Convert these measurements of length. Complete the table.

Metric	Imperial
30 cm	1 foot (12 inches)
	10 feet
	1 yard (3 feet)
	2 yards
1·8 m	feet

2. Convert these measurements of mass. Complete the table.

Metric	Imperial
450 grams	1 pound (16 ounces)
	0·5 pounds
900 grams	ounces
	3 pounds

3. Convert these measurements of capacity. Complete the table.

Metric	Imperial
about 500 ml	1 pint (20 fluid ounces)
about	10 fluid ounces
	about 4 pints
	about 8 pints (1 gallon)
4000 ml	

4 Mrs Bing is 6 feet tall. Mrs Bong is 174 cm tall. Who is taller? How much taller? Show how you worked it out.

★ **Challenge**

Put these animals in weight order from heaviest to lightest. You can use a calculator!

| 570 pounds | 24 stones | 200 kg | 2000 ounces |

---

1 Calculate the perimeter of these shapes.

a)

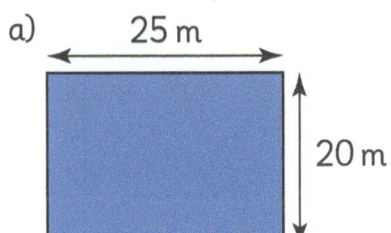
25 m
20 m

Perimeter = _____ m

b)

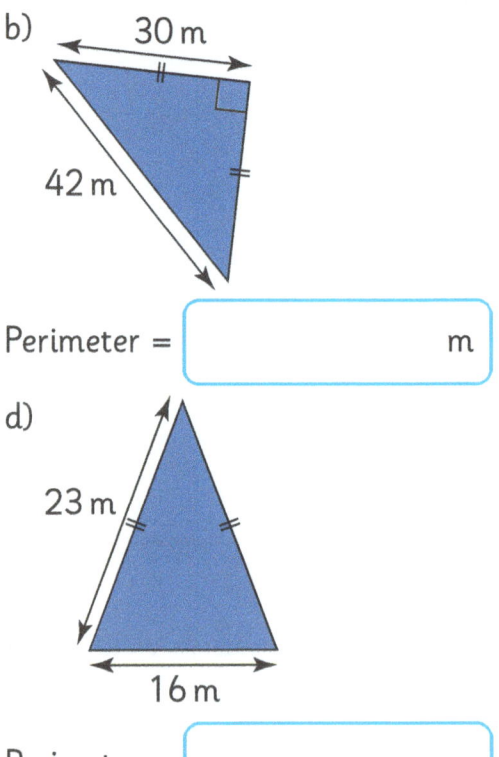
30 m
42 m

Perimeter = _____ m

c)

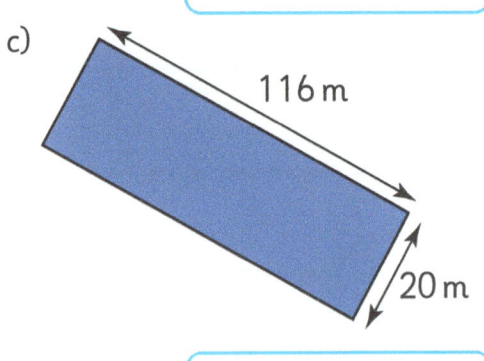
116 m
20 m

Perimeter = _____ m

d)

23 m
16 m

Perimeter = _____ m

2 You will need a ruler. Measure each side accurately and calculate the perimeter.

a)

Perimeter = _____ cm

b)

Perimeter = _____ cm

c)

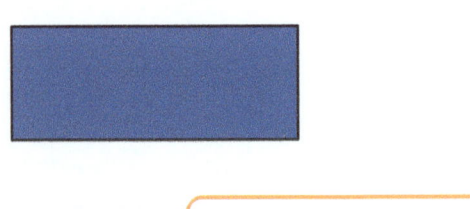

Perimeter = _____ cm

d)

Perimeter = _____ cm

3 Estimate the perimeter of each shape, then measure with a ruler and calculate the difference.

a)

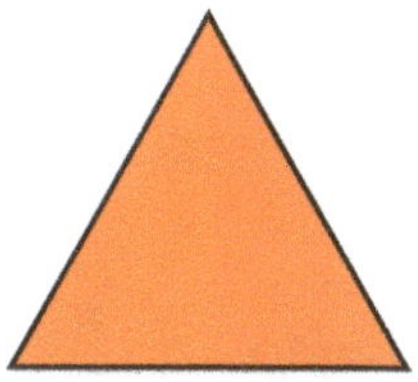

Estimate =

Actual =

Difference =

b)

Estimate =

Actual =

Difference =

c)

Estimate =

Actual =

Difference =

d)

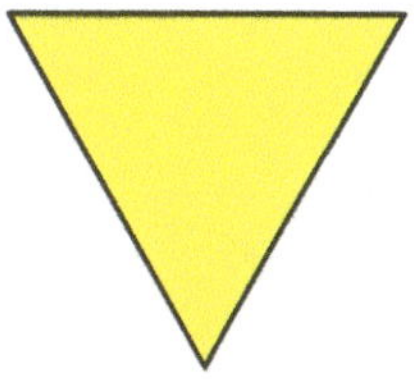

Estimate =

Actual =

Difference =

Draw as many four-sided shapes as possible, each with a perimeter of 160 mm. Name each shape.

9.7 Calculating the area of regular shapes

1 Measure, then calculate the area of these rectangles.

a)
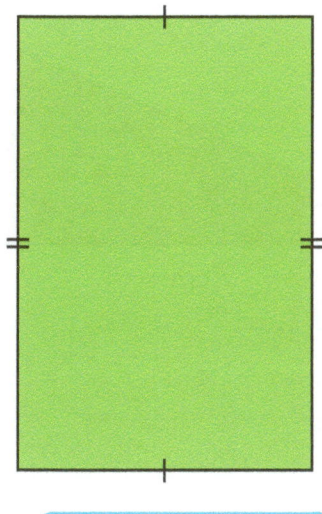

Area = [   ]

b)
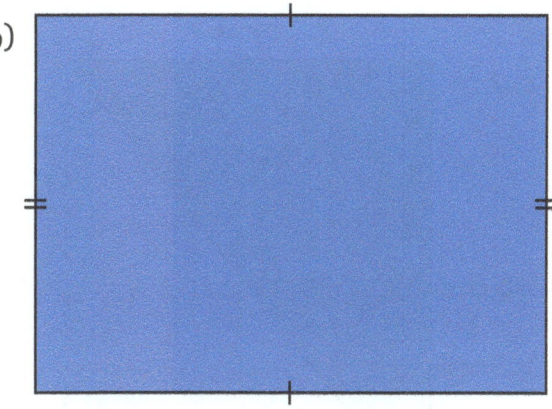

Area = [   ]

2 Draw triangles with areas of:

a) 6 cm^2 b) 8 cm^2 c) 10 cm^2

3 Draw in the lines to make each triangle into a rectangle. Calculate the area of the triangle.

a)

5 cm

6 cm

b)

3 cm

7 cm

$$= \frac{1}{2} \times \boxed{} \times \boxed{} = \boxed{} \text{ cm}^2$$

Area $= \frac{1}{2} \times$ base $\times$ height

$$= \frac{1}{2} \times \boxed{} \times \boxed{} = \boxed{} \text{ cm}^2$$

c)

4 cm

2 cm

d)

5 cm   3 cm

$$= \frac{1}{2} \times \boxed{} \times \boxed{} = \boxed{} \text{ cm}^2$$

$$= \frac{1}{2} \times \boxed{} \times \boxed{} = \boxed{} \text{ cm}^2$$

a) Draw a triangle that is **twice** the area of shape a.

b) Draw any shape that is **half** the area of shape b.

c) Draw a rectangle that is **half** the area of shape c and another rectangle that is a **quarter** the area of shape c.

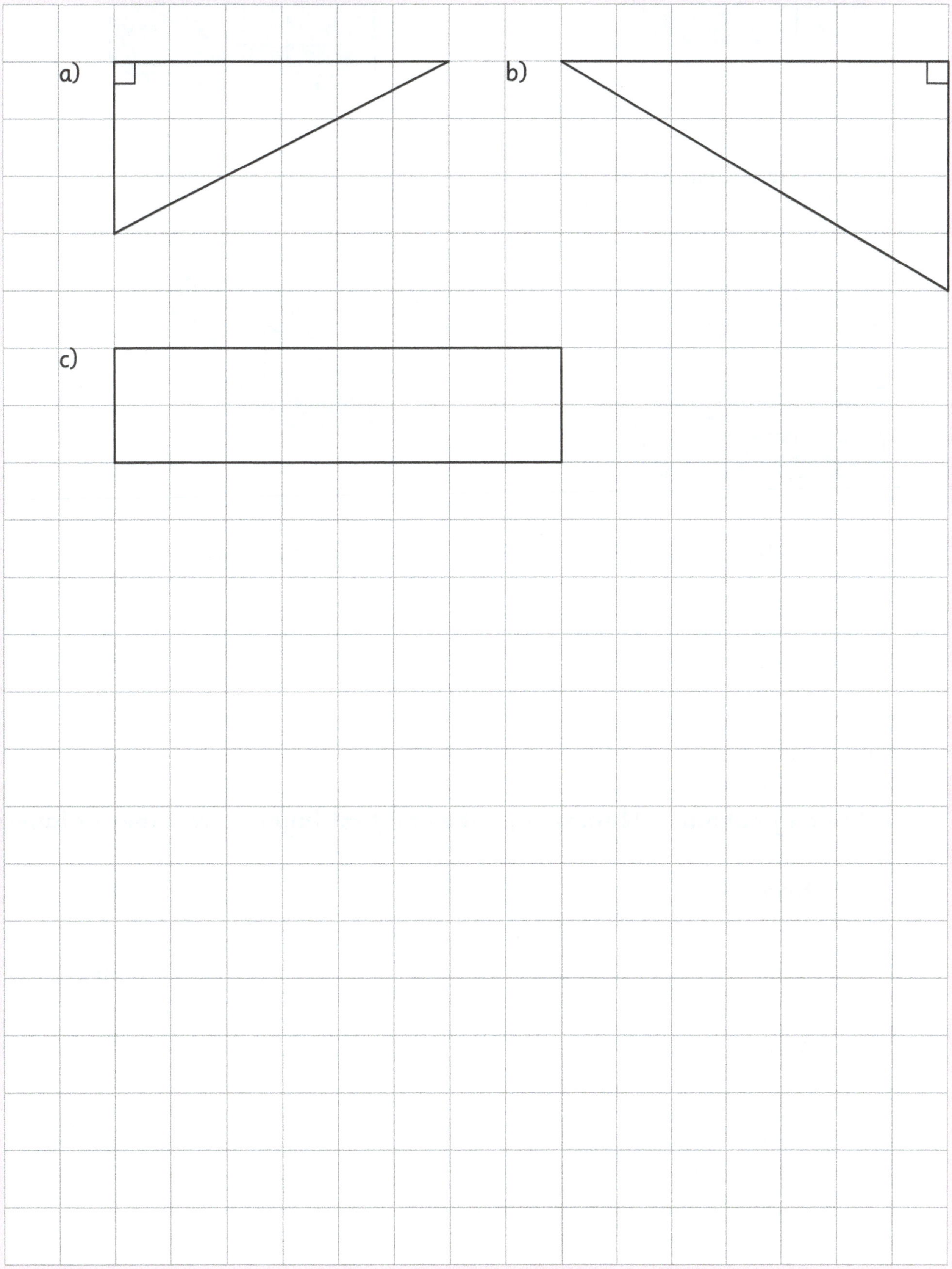

a)

b)

c)

1   Work out how many centimetre cubes are needed to make each cuboid.

a)

[ ] cubic centimetres

b)

[ ] cubic centimetres

c)

[ ] cubic centimetres

d)

[ ] cubic centimetres

2   Tyson builds cuboids with different volumes. Complete the table. The first one has been done for you.

Volume of cuboid	Number of cubes on first layer	Number of layers
18 cm³	9	2
20 cm³		5
25 cm³		1
28 cm³	7	
56 cm³		

3 We write cubic centimetres as cm³. Write down the volume of each of these cubes. The diagrams are not to scale so calculate by counting the smaller cubes that make up the shapes.

a)  Volume = ☐ cm³

b) 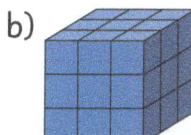 Volume = ☐ cm³

c) 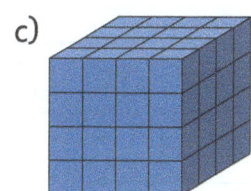 Volume = ☐ cm³

d) 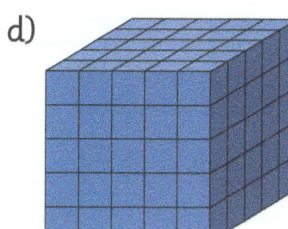 Volume = ☐ cm³

★ Challenge

Beth is building a model with 4 blocks. Her finished model has a volume of 170 cm³. What shape could block D be? Sketch your answer.

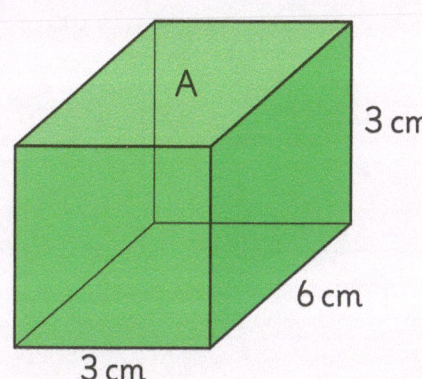

A
3 cm
6 cm
3 cm

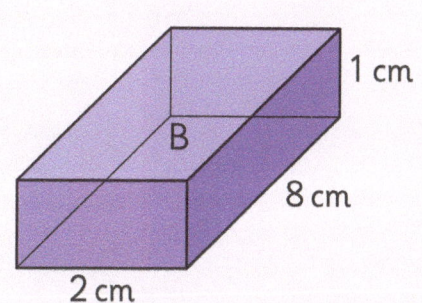

1 cm
B
8 cm
2 cm

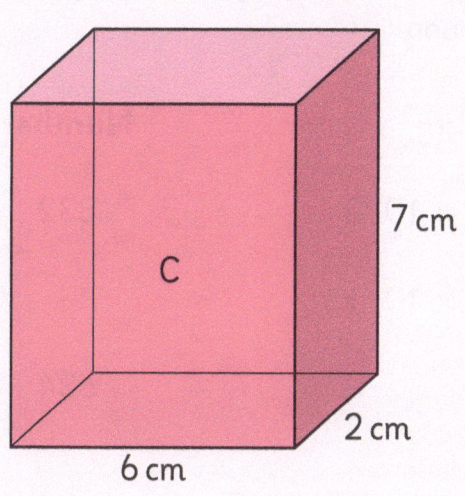

7 cm
C
2 cm
6 cm

The Chinese number system uses symbols called **characters** to represent each number.

0	zero	零	7	seven	七
1	one	一	8	eight	八
2	two	二	9	nine	九
3	three	三	10	ten	十
4	four	四	100	one hundred	百
5	five	五	1000	one thousand	千
6	six	六	10 000	ten thousand	万

1 The Chinese number system shows how many tens there are first. Complete the table. The first one is done for you.

Chinese Characters	Meaning	Number
八十五	8 × 10, 5	85
	3 × 10, 1	
		76
四十九		

2 The hundreds numbers are written by putting the number of hundreds first. Complete the table. The first one is done for you.

Chinese Characters	Meaning	Number
八百三十二	8 × 100, 3 × 10, 2	832
	9 × 100, 2 × 10, 1	
		634
二百九十五		

3 The thousands numbers are written by putting the number of thousands first. Complete the table. The first one is done for you.

Chinese Characters	Meaning	Number
四千三百五十六	4 × 1000, 3 × 100, 5 × 10, 6	4356
七千八百五十九		
		2163
	3 × 1000, 2 × 100, 6 × 10, 8	

4 Use the Chinese characters 一, 二, 三, 十, 百.

Use each character once to make 5-digit numbers. How many different 5-digit numbers can you make?

★ **Challenge**

Complete these calculations and give your answers in Chinese characters.

a) 三百四十四 + 二十二 =

b) 六百七十八 + 四 =

c) 六千三百五十一 + [   ] = 六千四百五十一

11.1 Exploring and extending number sequences

1 a) In the bakery, each cake has 6 chocolate buttons on top.
Complete the table to show the pattern.

No. of cakes (c)	1	2	3	4	5
No. of buttons (b)	6				

b) Write the pattern you notice in words.

c) The number of buttons is [　　　] × the number of cakes.

d) Complete the formula b = [　　　] × [　　　]

2 a) In the garden centre each tub has 8 flowers in it.
Complete the table to show the pattern.

No. of tubs (t)	1	2	3	4	5
No. of flowers (f)	8				

b) The number of flowers is [　　　] × [　　　]

c) Write the formula in symbols [　　　]

d) Use the formula to work out how many flowers there would be if there were 12 tubs.

3 a) Each heptagon is made of 7 matchsticks. Complete the
   table to show the pattern.

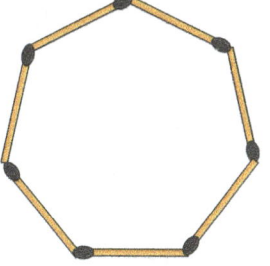

No. of heptagons (h)	1	2	3	4	5
No. of matchsticks (m)	7				

b) Write the formula.

c) Use the formula to work out how many matchsticks there would be if there were
   20 heptagons.

★ Challenge

Imogen's dog ate part of her homework. Complete the table.

Cards (C)		9	10
Stamps (S)		36	40

Cards (c)									9	10
Stamps (s)									36	40

b) Write the formula here:

12.1 Solving equations using mathematical rules

1 Use one of these symbols **< > = ≠** to make each statement true.
Each symbol must be used once only.

a) $156 + 4$ ☐ $160 \div 4$

b) $156 + 40$ ☐ $200 - 4$

c) $686 - 120$ ☐ $526 + 50$

d) $567 + 120$ ☐ $526 - 50$

2 Choose one of these calculations to balance each equation. Each calculation must be used at least once.

| 24×1 | 12×1 | 6×2 | 6×4 |

a) $4 \times 6 =$ ☐

b) $3 \times 8 =$ ☐

c) $48 \div 2 =$ ☐

d) $4 \times 3 =$ ☐

e) $36 \div 3 =$ ☐

f) $\frac{1}{2}$ of $48 =$ ☐

3 Find the missing number in each equation.

a) ☐ $- 4 = 3 \times 5$

b) $15 +$ ☐ $= 25 - 6$

c) $25 \times 2 =$ ☐ $\times 10$

d) $50 \div 10 = 60 -$ ☐

e) $55 \div 11 =$ ☐ $\div 5$

f) ☐ $+ 50 = \frac{3}{4}$ of 100

Each of these shapes represents a number. Work out the value of each shape.

 = 10

 = 13

 = 9

= 15

 = 14

Jottings

13.1 Naming and sorting shapes

1 A regular polygon has congruent sides and angles. What does this mean?

2 Are these shapes **regular** or **irregular**? Write each letter in the correct box.

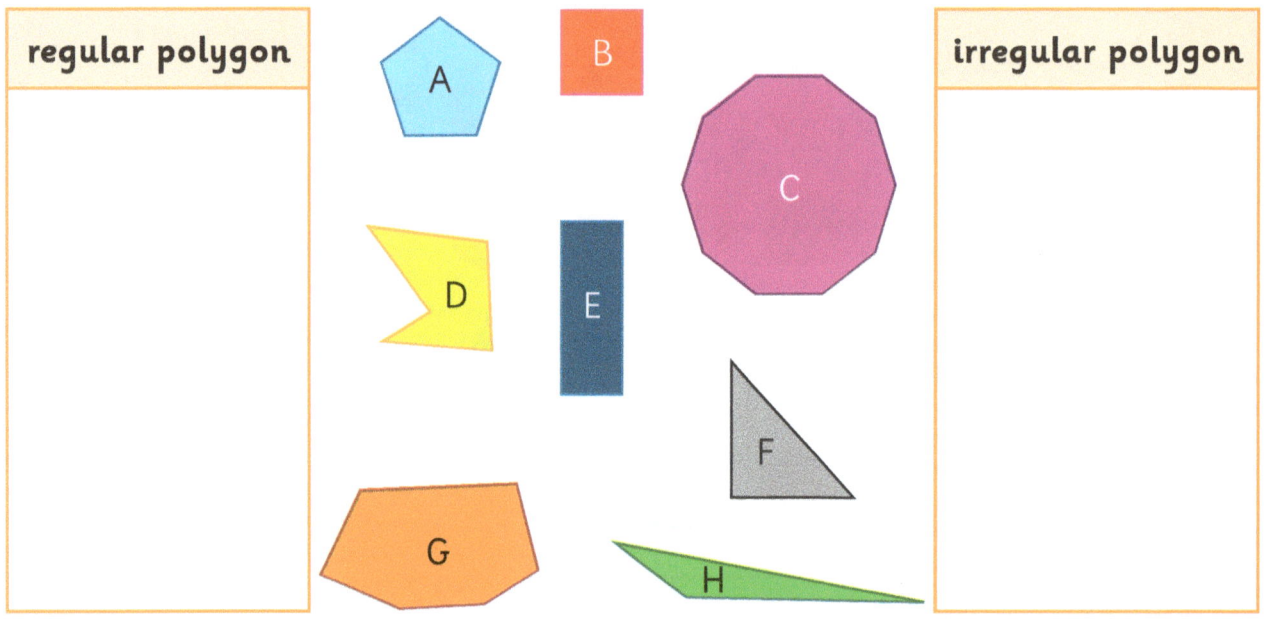

regular polygon		irregular polygon

3 Draw and then name these shapes. Make sure to say if the shapes are **regular** or **irregular**.

a) I have three straight sides.

   My sides are different lengths.

   I have three vertices.

b) I have four straight sides.

   Two of my sides are different lengths.

   I have four vertices.

c) I have six straight sides.

My sides are the same length.

I have six vertices.

d) I have five straight sides.

My sides are different lengths.

I have five vertices.

⭐ **Challenge**

a) Look at these shapes. Decide how to sort them and put them into two groups. Label your groups to show your criteria.

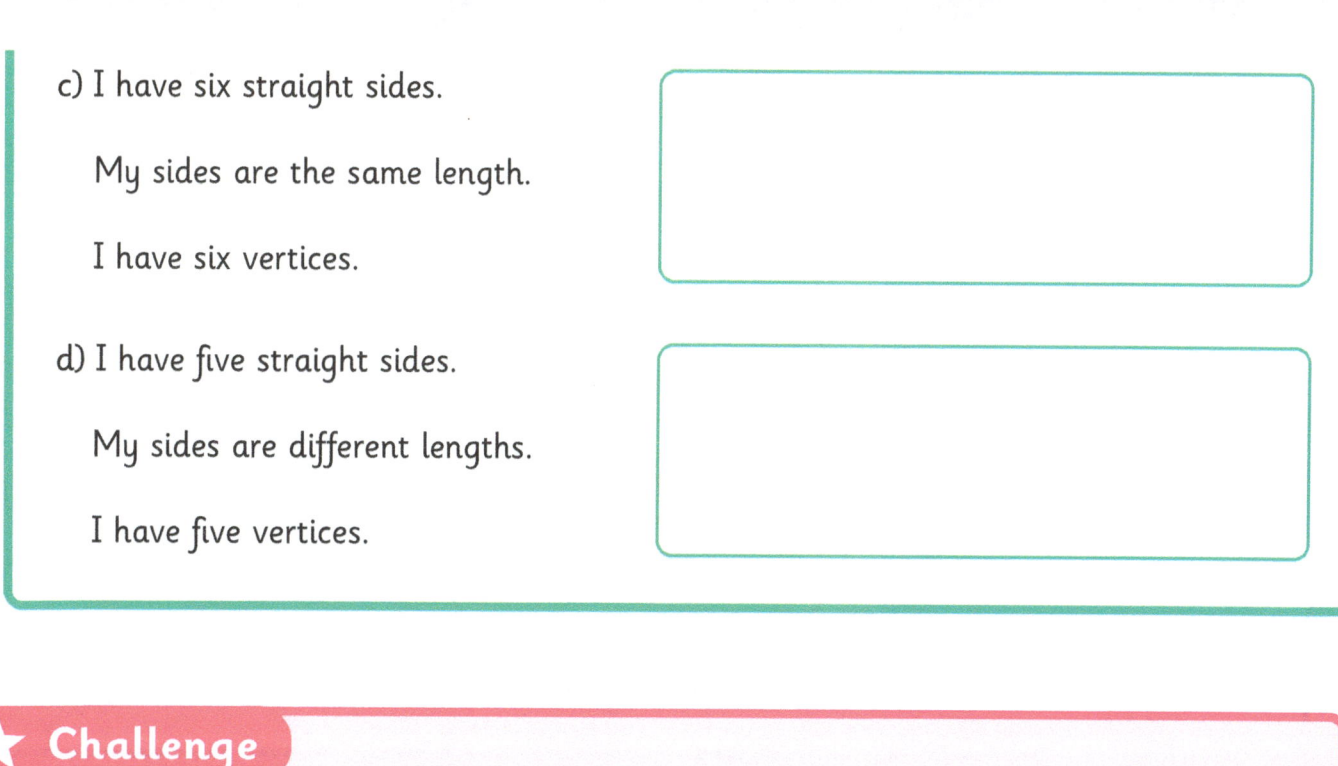

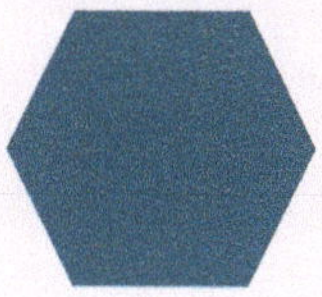

Group 1

Label:

Group 2

Label:

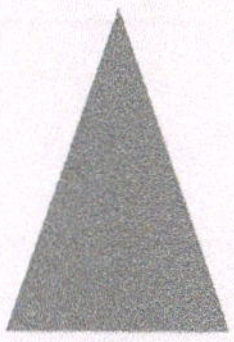

13.2 Describing and drawing circles

1) Label the diameter, radius and circumference of this circle.

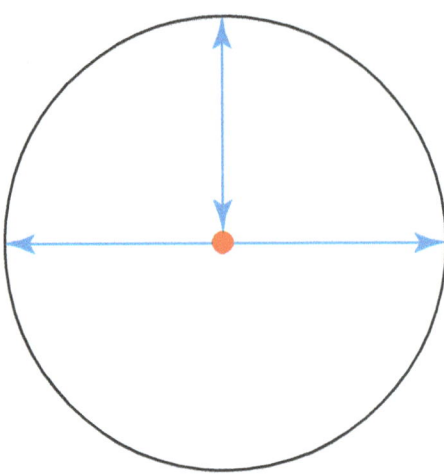

2) a) Use a pair of compasses to draw a circle with a radius of 4 cm and a circle with a diameter of 6 cm. Your circles can overlap.

b) Measure the diameter and radius of both of your circles.

Circle 1 diameter =

Circle 1 radius =

Circle 2 diameter =

Circle 2 radius =

3 Complete the table.

Shape	Radius	Diameter	Approximate circumference
A	11 cm		
B		44 cm	
C			120 mm
D		38 cm	
E	18 m		

★ **Challenge**

Create a design using four circles. Use a pair of compasses. First, draw a circle that has a radius of 2 cm. Now add a circle with a diameter of 8 cm. Then add a circle with an approximate circumference of 30 cm. Finally, add a circle of your choosing.

13.3 Constructing 3D objects

1 Match each skeleton to the correct name.

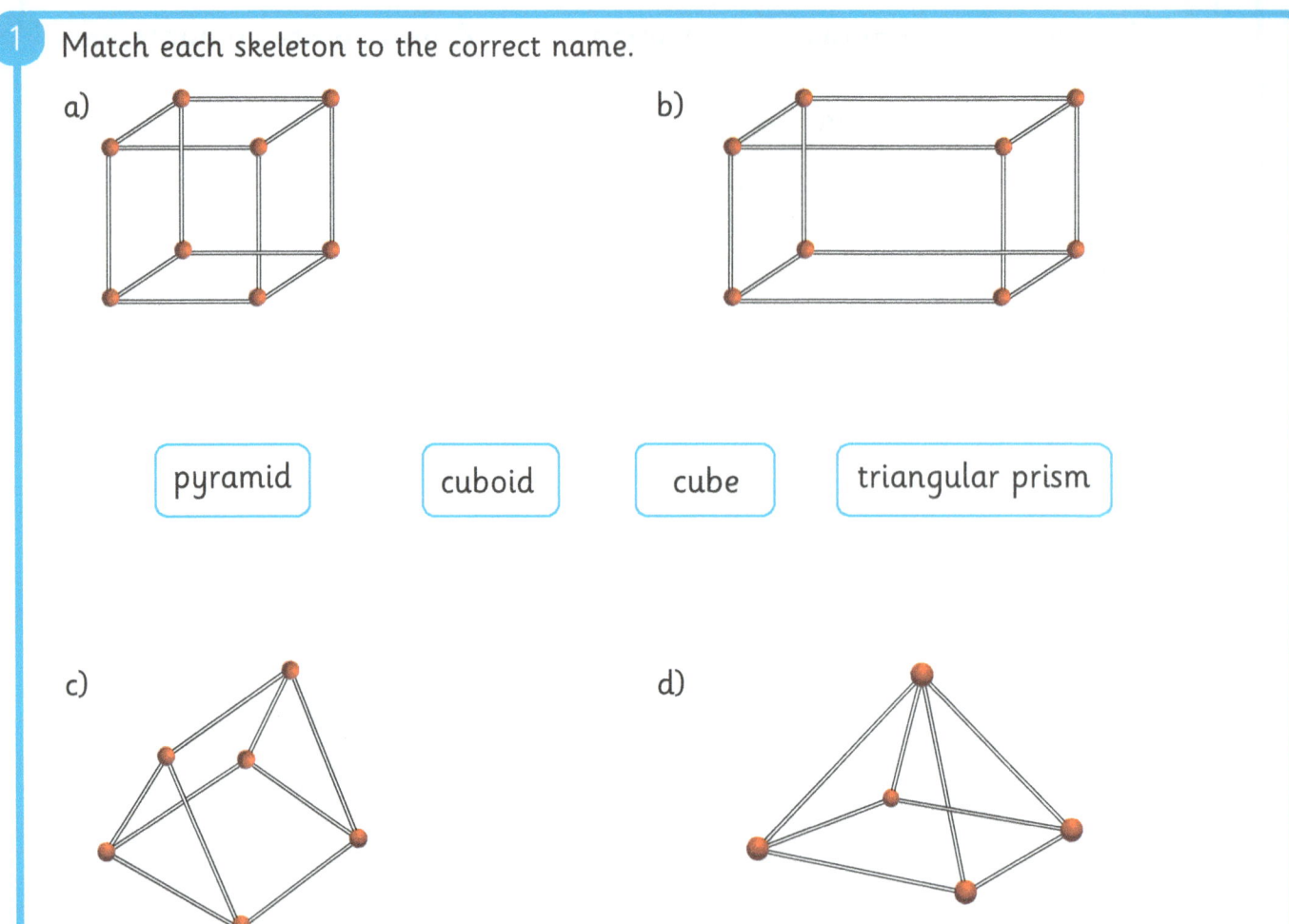

a)

b)

| pyramid | cuboid | cube | triangular prism |

c)

d)

2 Find examples of 3D objects to help you complete the table. Answer yes or no. The first one is done for you.

Object	More than two edges meeting at a vertex	Can have a square face
	Yes	Yes

3 Draw in all the missing edges in each of these and write the name of the 3D object underneath.

a)

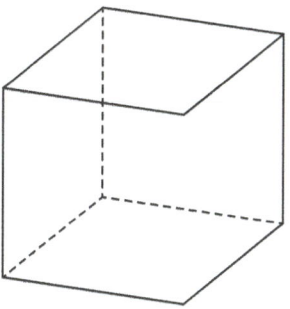

b)

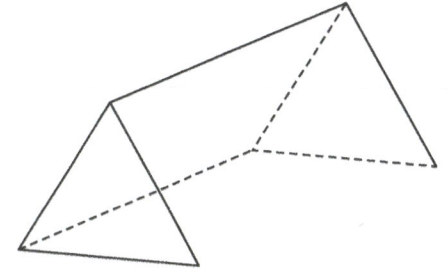

c)

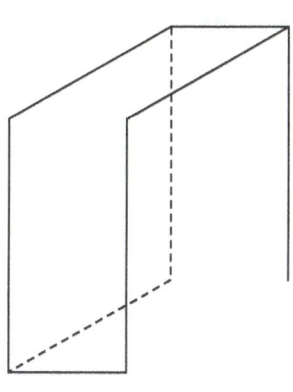

d)

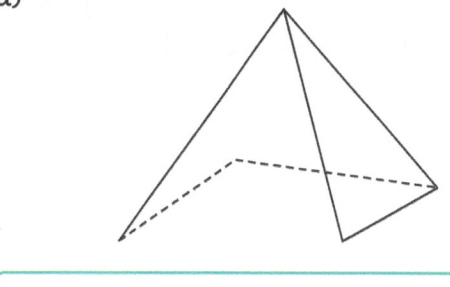

You will need scissors, straws and sticky tape or modelling clay.

a) Use four long straws and eight short straws to build a cuboid. Sketch it here:

How many vertices does it have?

b) Remove two short straws and one long straw from your cuboid and create a triangular prism from the remaining skeleton. Sketch it here:

How many vertices does it have?

c) Remove three short straws from your triangular prism and create a triangular based pyramid from the remaining skeleton. Sketch it here:

How many vertices does it have?

13.4 Constructing nets

1

You will need centimetre squared paper.

a) Copy this net onto the centimetre squared paper.

I predict this net will fold into a 3D object called a

b) Cut out your net, fold and create the 3D object.

My object is a

2

a) Copy this net onto the centimetre squared paper.

I predict this net will fold into a 3D object called a

b) Cut out your net, fold and create the 3D object.

My object is a

3

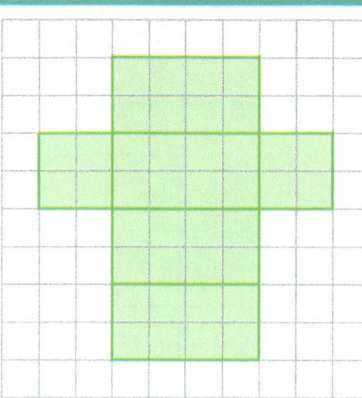

a) Copy this net onto the centimetre squared paper.

I predict this net will fold into a 3D object called a [ ]

b) Cut out your net, fold and create the 3D object.

My object is a [ ]

4

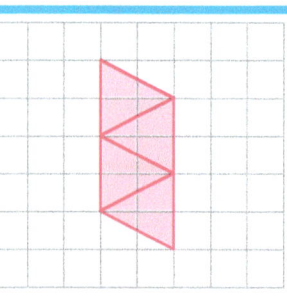

a) Copy this net onto the centimetre squared paper.

I predict this net will fold into a 3D object called a [ ]

b) Cut out your net, fold and create the 3D object.

My object is a [ ]

★ Challenge

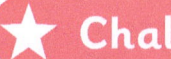

You will need:

- Squared paper if you run out of space here!
- A ruler

Make as many different nets of a triangular prism as you can. Draw them here:

14.1 Identifying and sorting angles

1 Match these angles to the correct name and description.

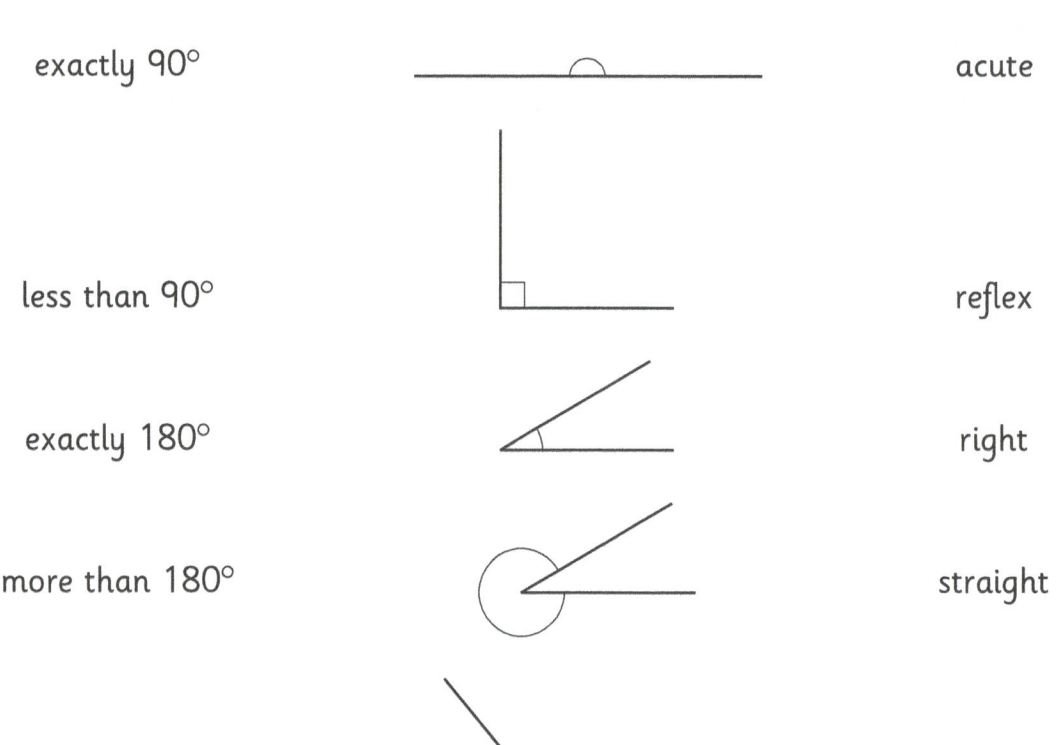

exactly 90°   acute

less than 90°   reflex

exactly 180°   right

more than 180°   straight

more than 90°,   obtuse
less than 180°

2 Sort these angles and complete the table.
One has been done for you.

	Acute	Obtuse	Reflex
a)			

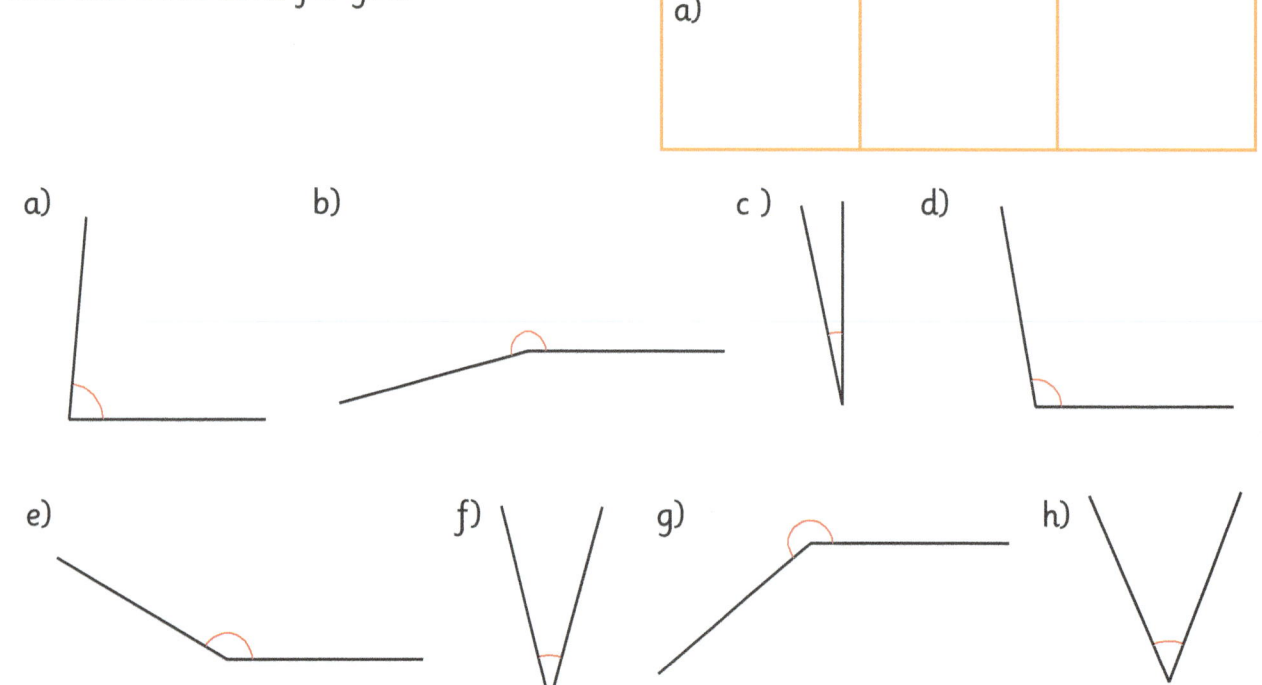

3 Correct the students' work. Tick the correctly identified angles and put a cross at the incorrect ones.

Cole (reflex angles):

Margo (obtuse angles):

Alissa (acute angles):

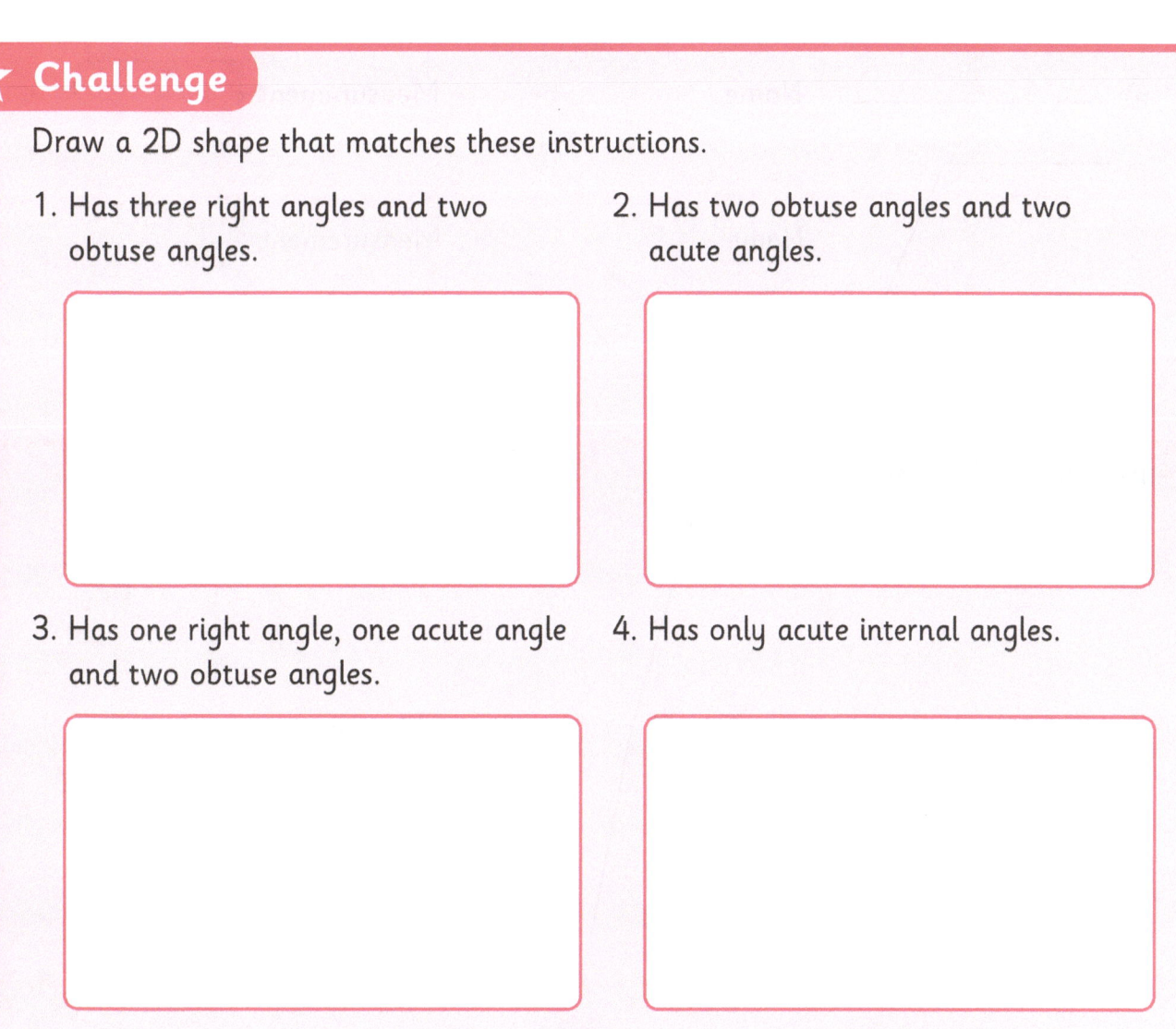

Draw a 2D shape that matches these instructions.

1. Has three right angles and two obtuse angles.

2. Has two obtuse angles and two acute angles.

3. Has one right angle, one acute angle and two obtuse angles.

4. Has only acute internal angles.

14.2 Measuring and drawing angles

1 Measure and name these angles. The first is done for you.

a)

Name [right] Measurement [90°]

b)

Name [ ] Measurement [ ]

c)

Name [ ] Measurement [ ]

d)

Name [ ] Measurement [ ]

e)

Name [ ] Measurement [ ]

f)

Name [ ] Measurement [ ]

2 Estimate and measure:

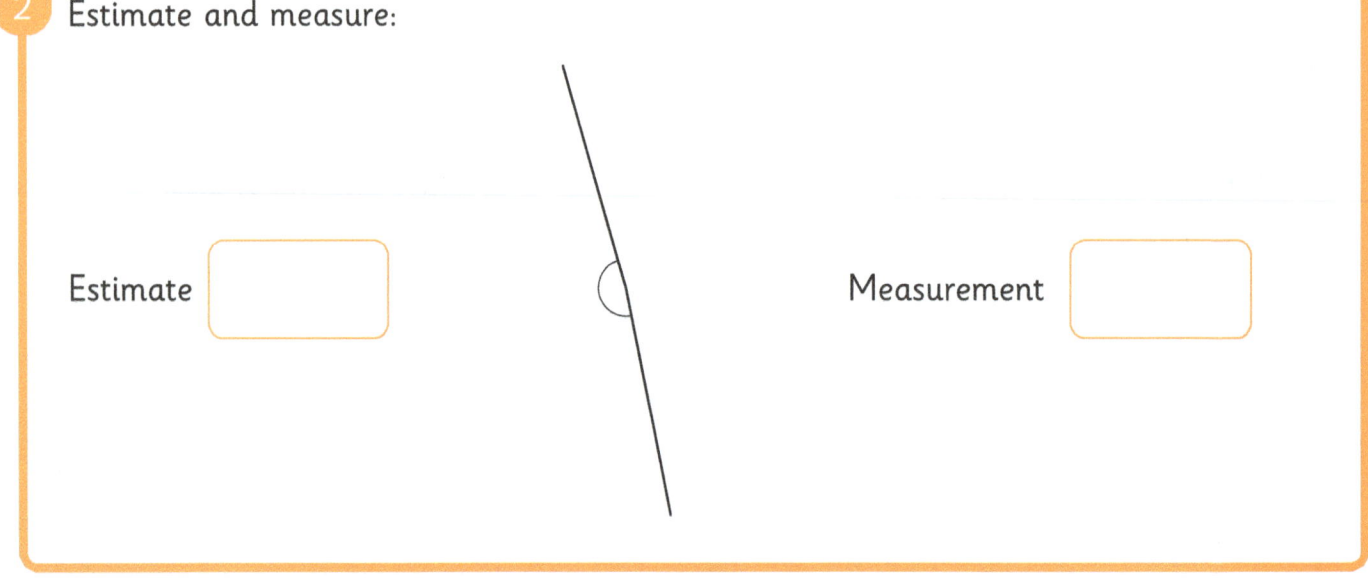

Estimate [ ] Measurement [ ]

3 Draw and label these angles using a protractor on the line below.

a) 56°

b) 156°

c) 256°

⭐ **Challenge**

Find and write the names of 5 different four-sided shapes with inside angles that total 360°.

Now draw them accurately below.

14.3 Finding missing angles

1 Match the definition to the name.

Two angles that add up to 90°

complementary

supplementary

Two angles that add up to 180°

2 Calculate the missing complementary angles.

a)

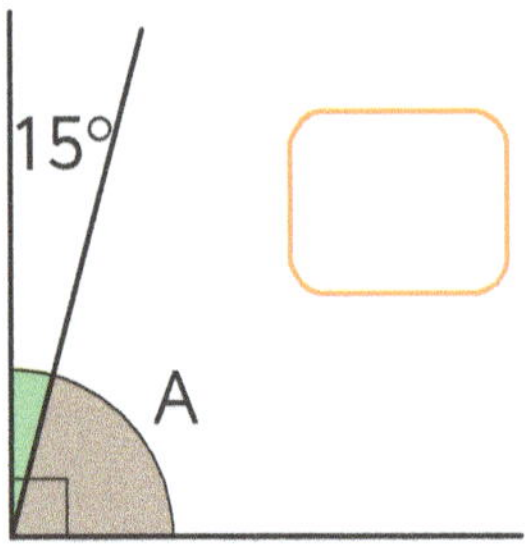

15°

A

b)

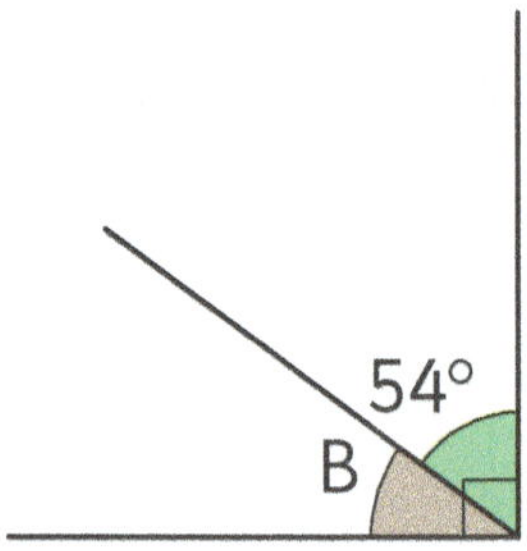

54°

B

c)

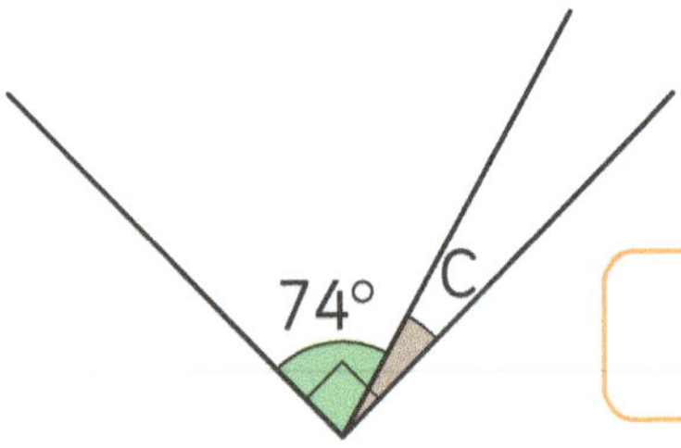

74° C

d)

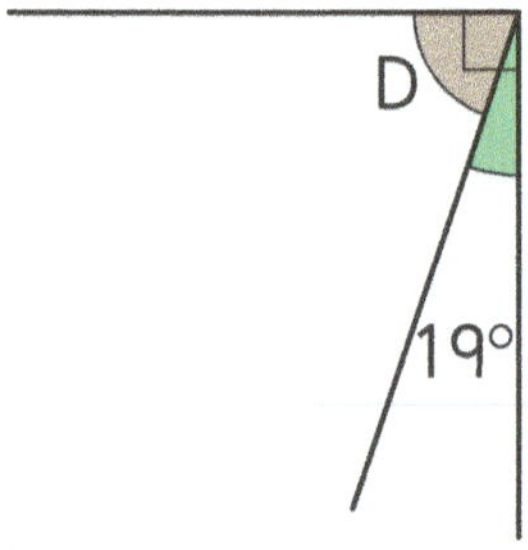

D

19°

e)

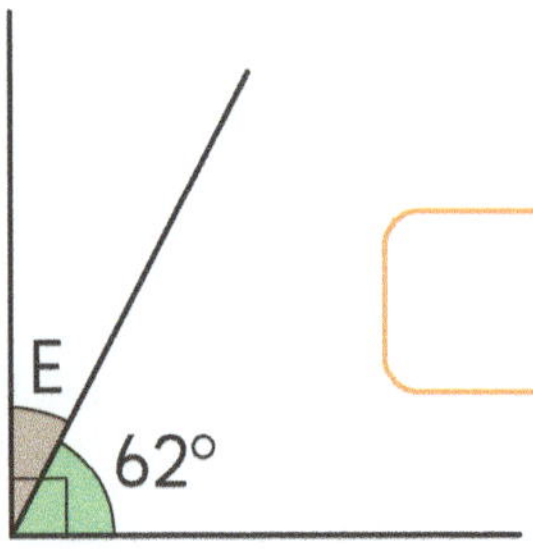

E

62°

f)

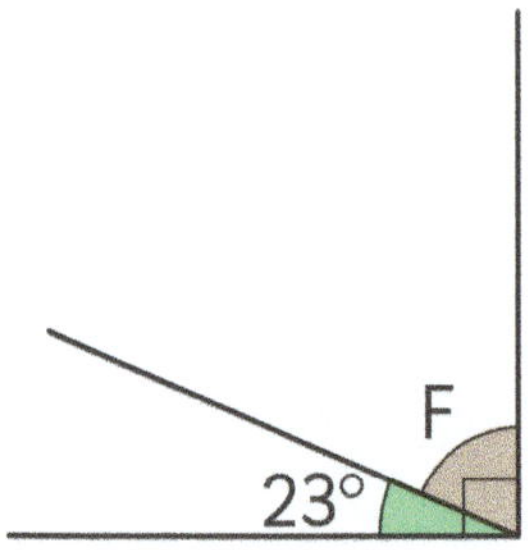

F

23°

3 Calculate the missing complementary angles.

a)

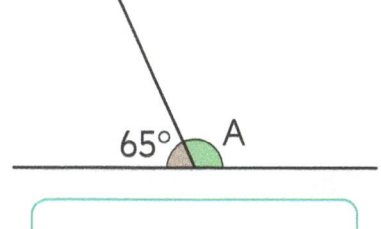

65° A

b)

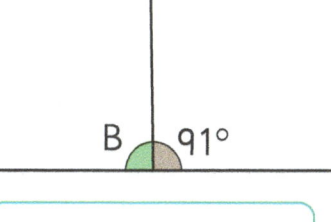

B 91°

c)

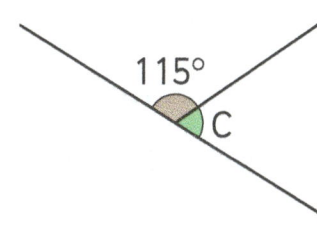

115° C

d)

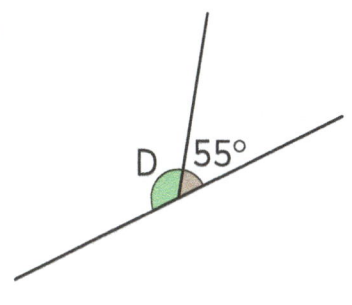

D 55°

e)

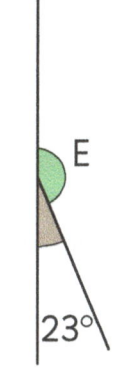

E

23°

f)

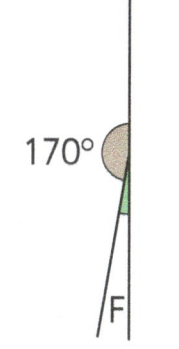

170°

F

4 Measure and find the missing angles.

a)

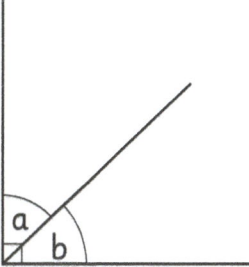

a

b

b)

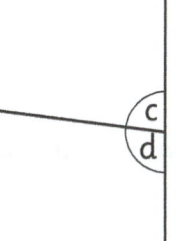

c
d

c)

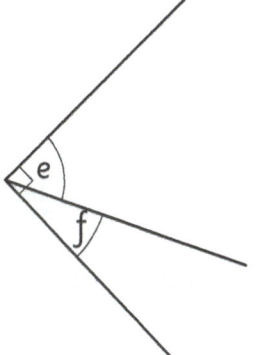

e

f

Measure angle a =

Calculate angle b =

Measure angle c =

Calculate angle d =

Measure angle e =

Calculate angle f =

This is a Frayer model for right angles:

Definition	Characteristics
An angle that measures 90°	The angle created when two straight lines are perpendicular to each other.
Example	Non example

Create a Frayer model for complementary angles.

Definition	Characteristics
Example	Non example

Create a Frayer model for supplementary angles.

Definition	Characteristics
Example	Non example

1 Complete the compass rose with 3-figure bearings.

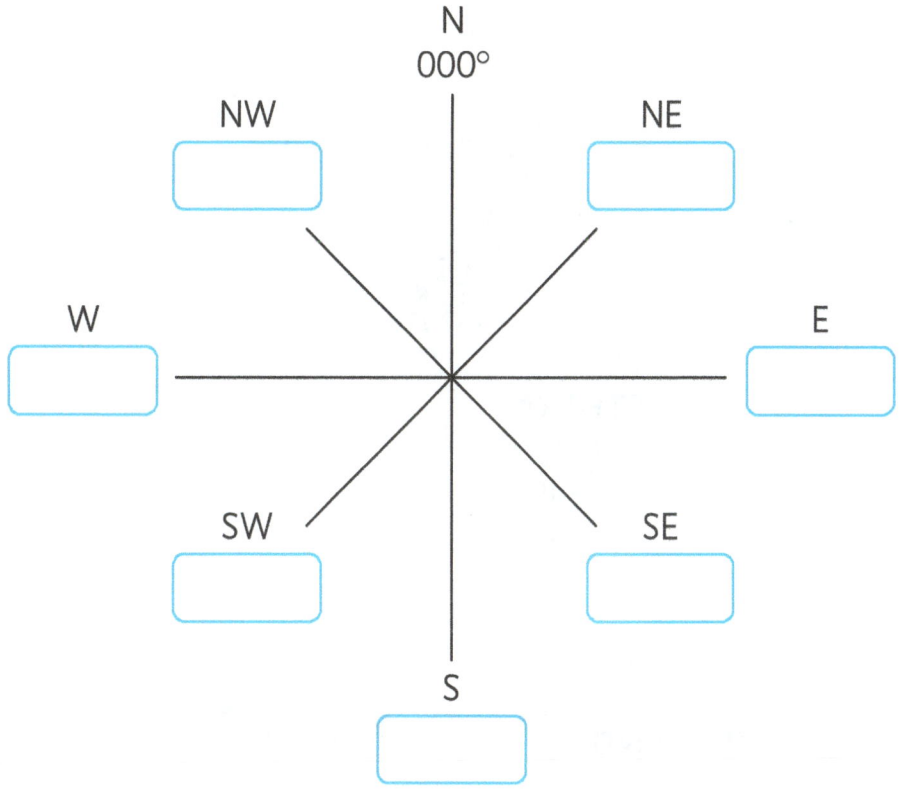

2 Complete the table by ticking the correct answer.

	True	False	Do not know
Bearings must always have 3 digits			
It is possible to have bearings over 359°			
Bearings are used in map reading			
Bearings start at either North, South, East or West			

Write your own fact about bearings here. Ask a partner to answer true / false / do not know. Are they correct?

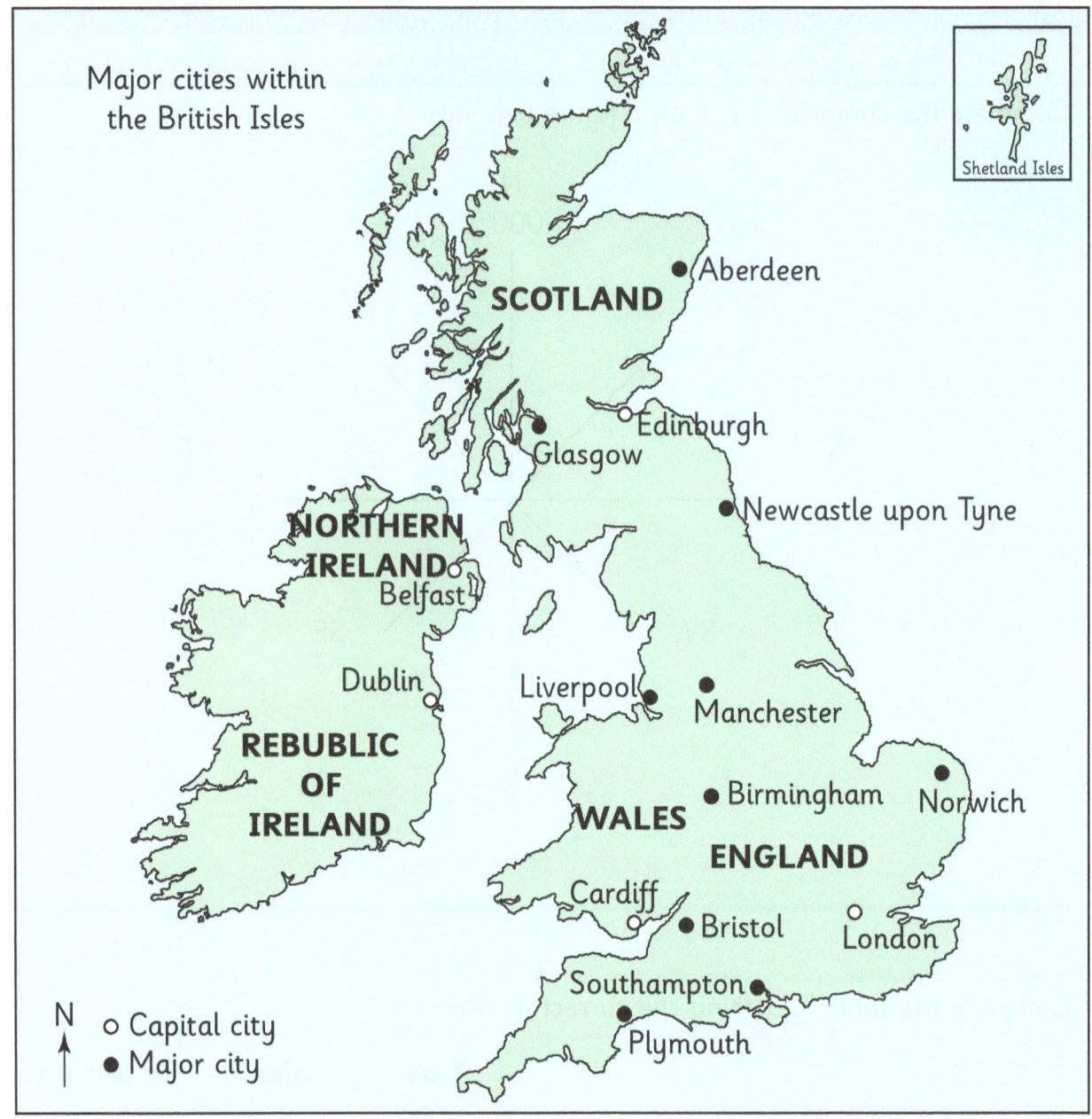

Major cities within the British Isles

You will need a protractor or angle measure. Ms Keith is a pilot. She needs to use bearings to know which direction to fly in. Complete the table to show bearings from these airports.

From	To	Bearing
Glasgow	Aberdeen	
Aberdeen	Newcastle upon Tyne	
Edinburgh	Belfast	
London	Cardiff	
Bristol	Glasgow	

Create an imaginary map below. It can be anywhere in the world! Put six landmarks on your map and mark them A, B, C, D, E and F. Create a tour for a visitor to your area. Give them instructions below your map.

From	To	Bearing
A	B	
B	C	
C	D	
D	E	
E	F	

14.5 Reading coordinates

1 Write down the coordinates of the vertices of each shape.

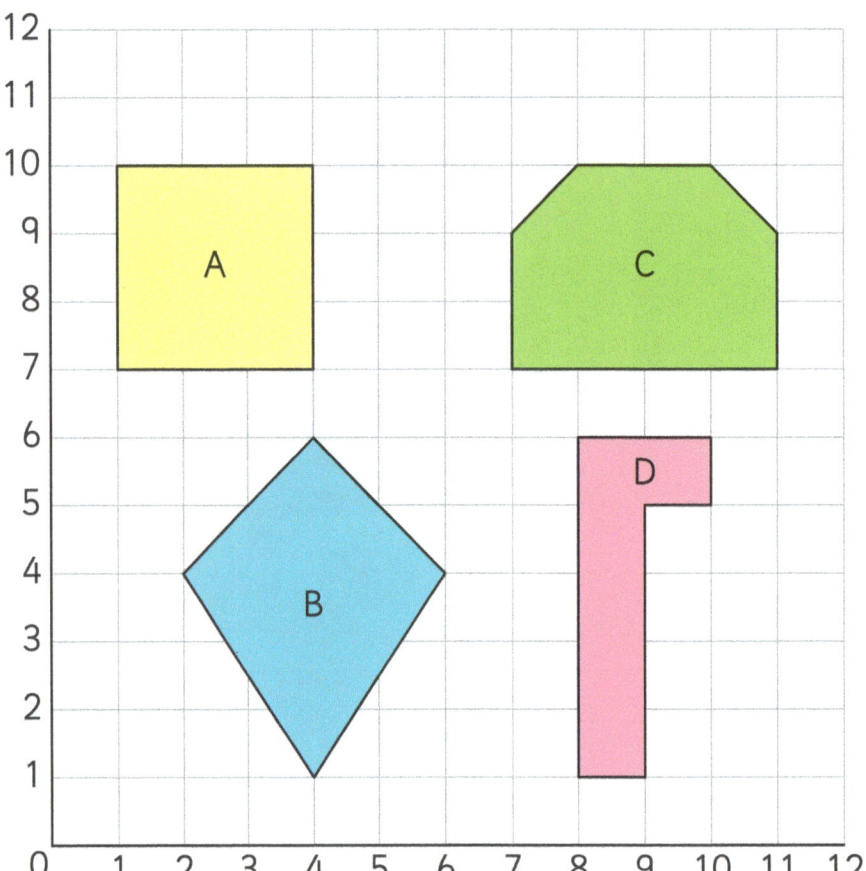

a)

b)

c)

d)

2 a) Plot the following coordinates. Join the dots to reveal the image.

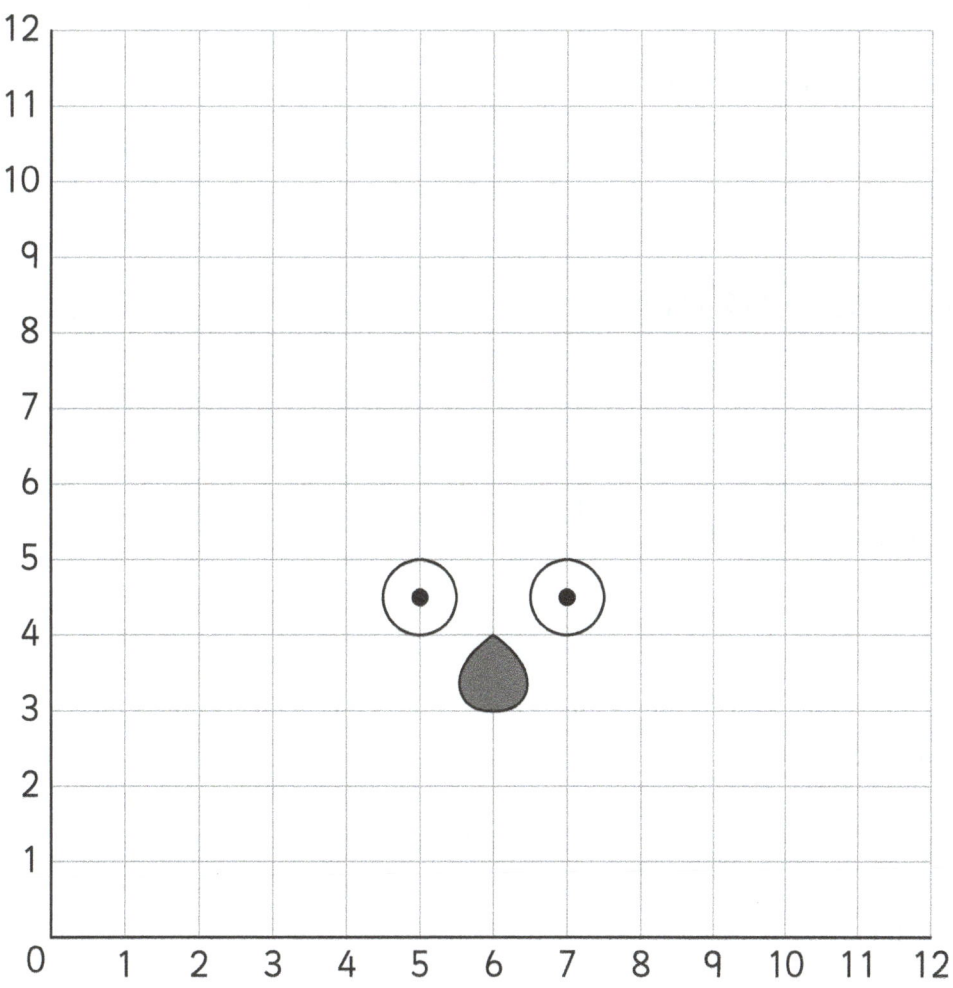

(4,11), (3,9), (3,7), (2,6), (2,2), (5,1), (7,1), (10,2), (10,6), (9,7),
(9,9), (8,11), (7,9), (7,7), (5,7), (5,9)

b) Add in six whiskers and write the coordinates here:

3 Identify the missing coordinates in these shapes. Draw them on the grid.

a) Square (1,6), (1,10) (5, 10)

b) Rectangle (12,5), (12,7) (6,5)

c) Square (2,1), (10,1) (10,9)

d) Rectangle (2,2), (2,4) (8,4)

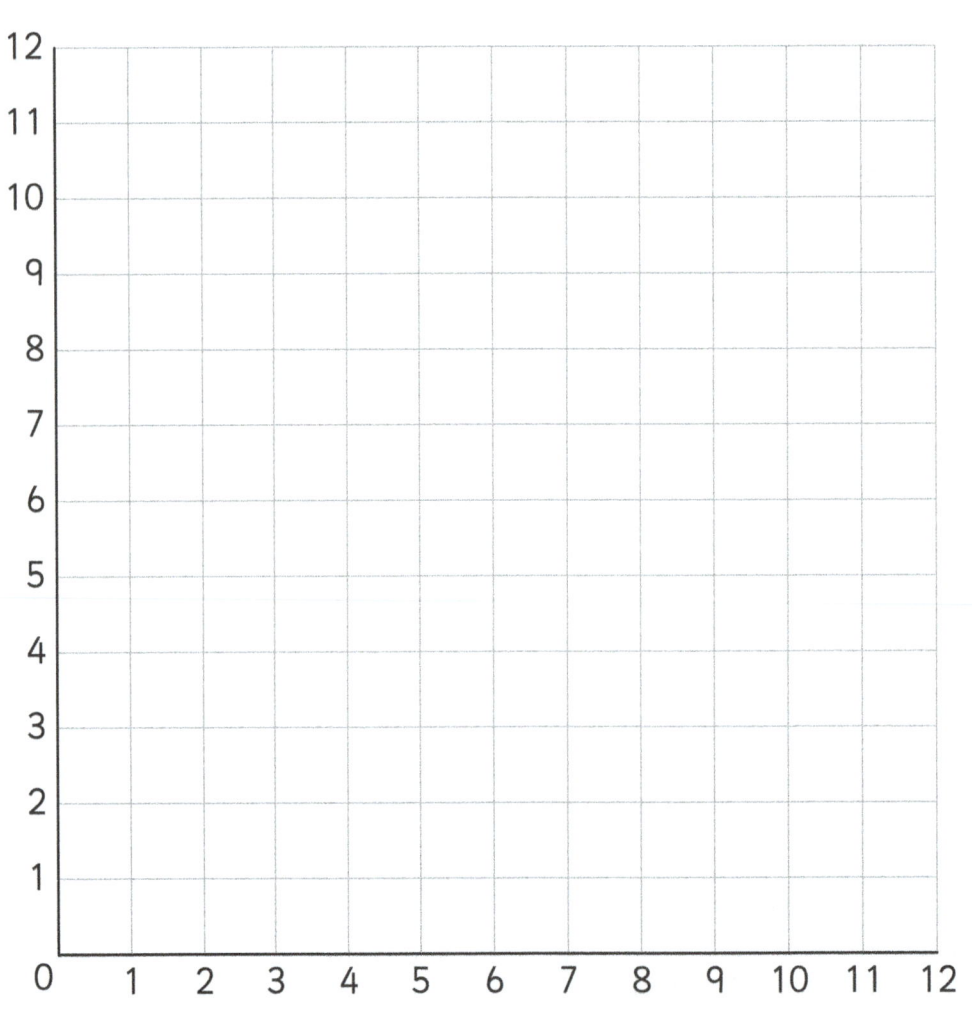

You are going to create an image using the first letter of your first name and the first letter of your surname. You must try to fill the space so your letters will overlap. An example is shown here using K and H:

(1,0), (2,0), (6,0), (7,0), (7,11), (6,11), (6,6), (6,5), (2,11), (1,11) (2,6), (2,5)

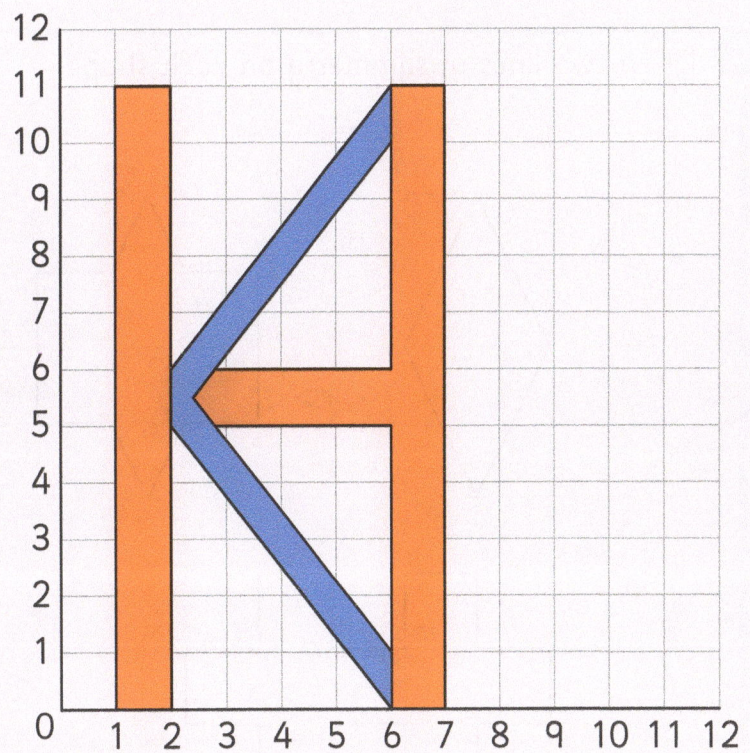

Colour your design and write the coordinates in the box:

14.6 Line symmetry

1 Draw two lines of symmetry on each shape.

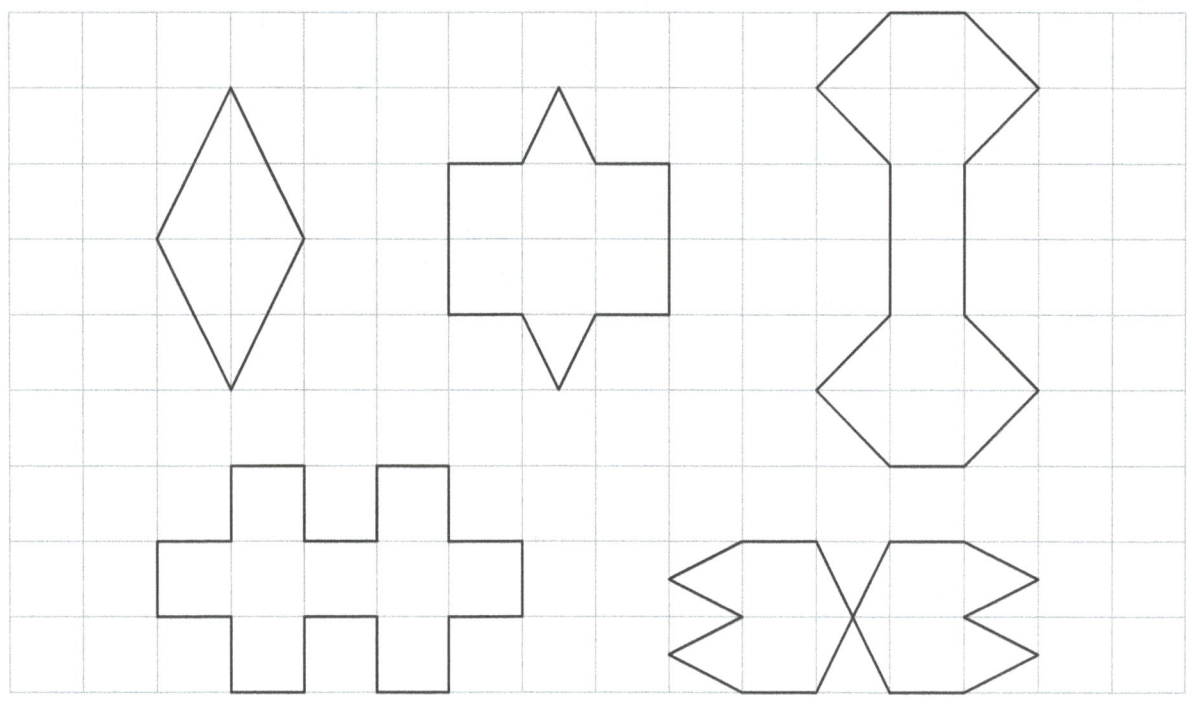

2 You must use three different colours. Colour the squares so that each one has at least two lines of symmetry. An example is shown below. You could use a mirror to check your designs.

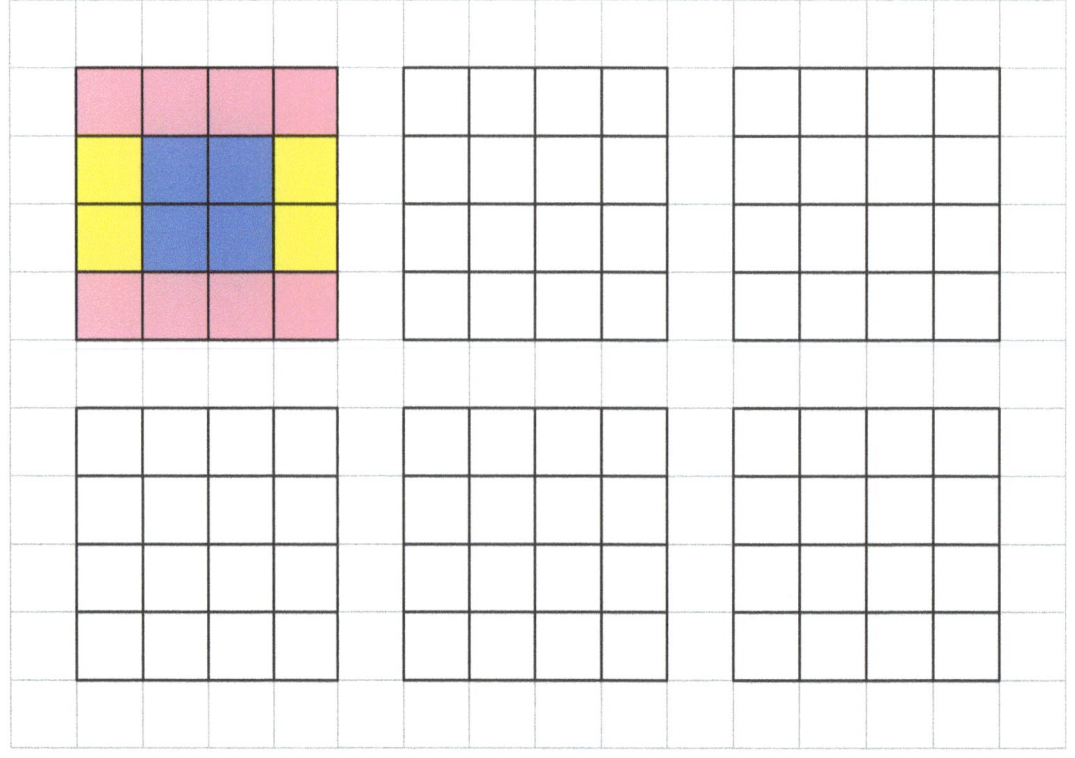

3 Look at these designs. Complete the table.

a)   b)   c)   d)

e)   f)   g)   h)

No lines of symmetry	One line of symmetry	Two or more lines of symmetry

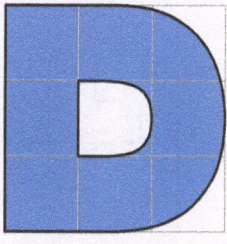

1 Complete these shapes so that they each have two lines of symmetry.

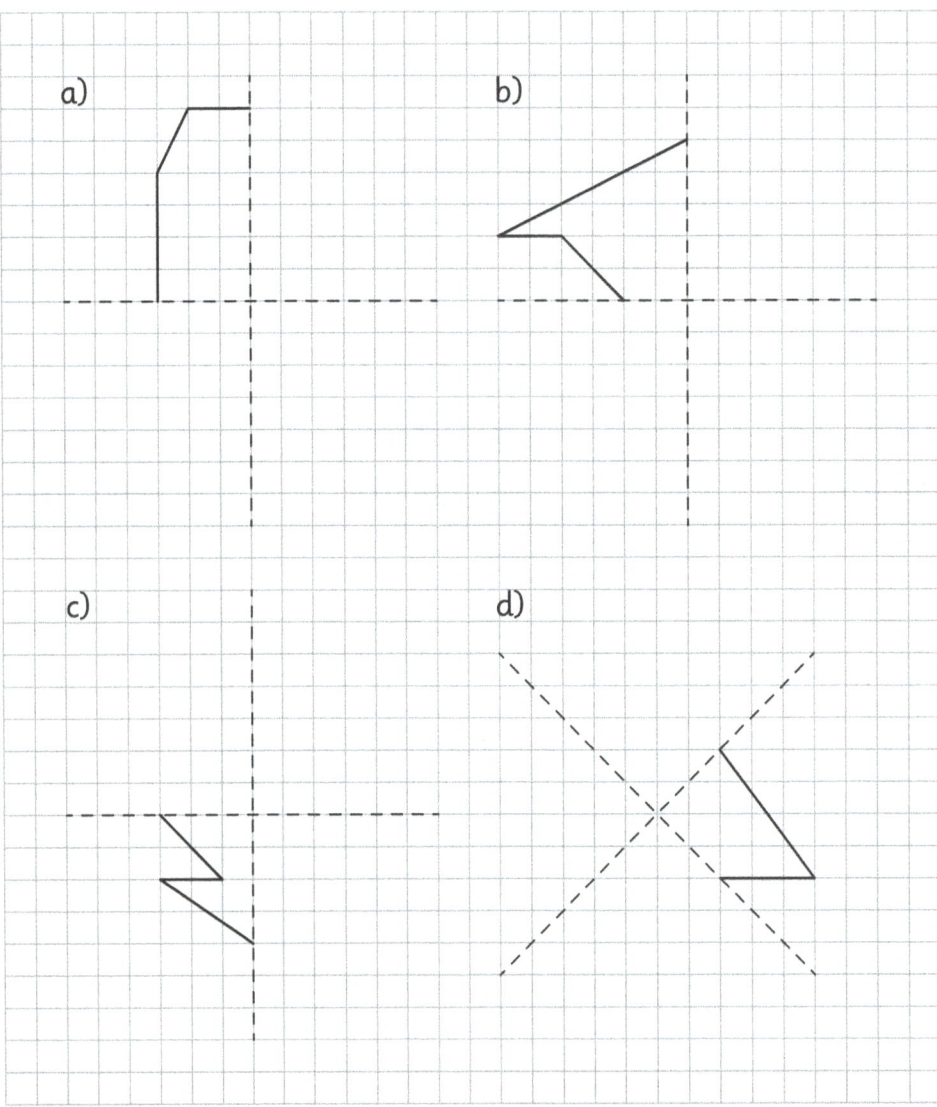

a)

b)

c)

d)

2 You will need to use red, white and blue. Colour each flag in the correct colours. Underneath each flag write the number of lines of symmetry it has.

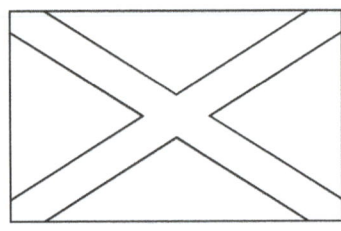

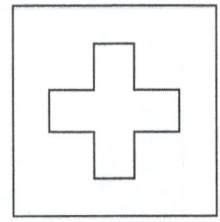

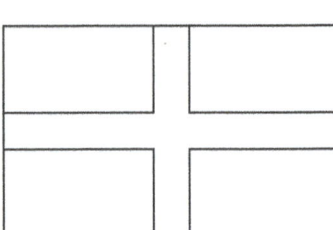

Scotland

Switzerland

England

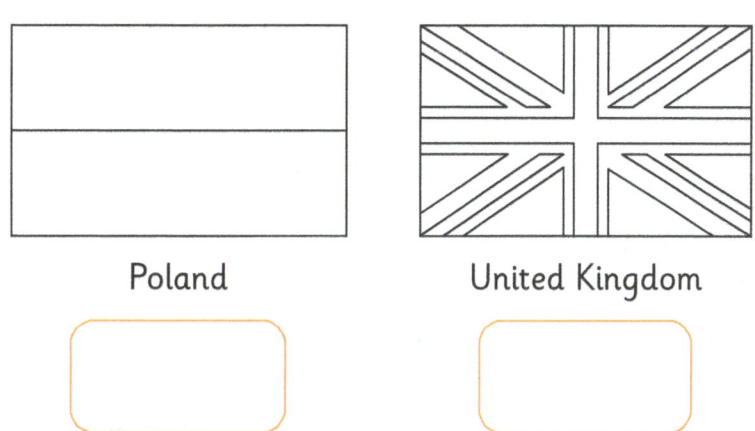

Poland
United Kingdom

3 Decide if these statements are **always true**, **sometimes true** or **never true**. Circle the answer in the box.

Triangles have only one line of symmetry. always sometimes never	Squares have four lines of symmetry. always sometimes never
A circle has only one line of symmetry always sometimes never	Regular pentagons have six lines of symmetry. always sometimes never
The number of lines of symmetry and the number of sides in regular polygons is the same. always sometimes never	

Challenge

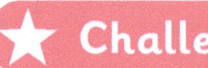

Colour this pattern using five colours so that it has two lines of symmetry.

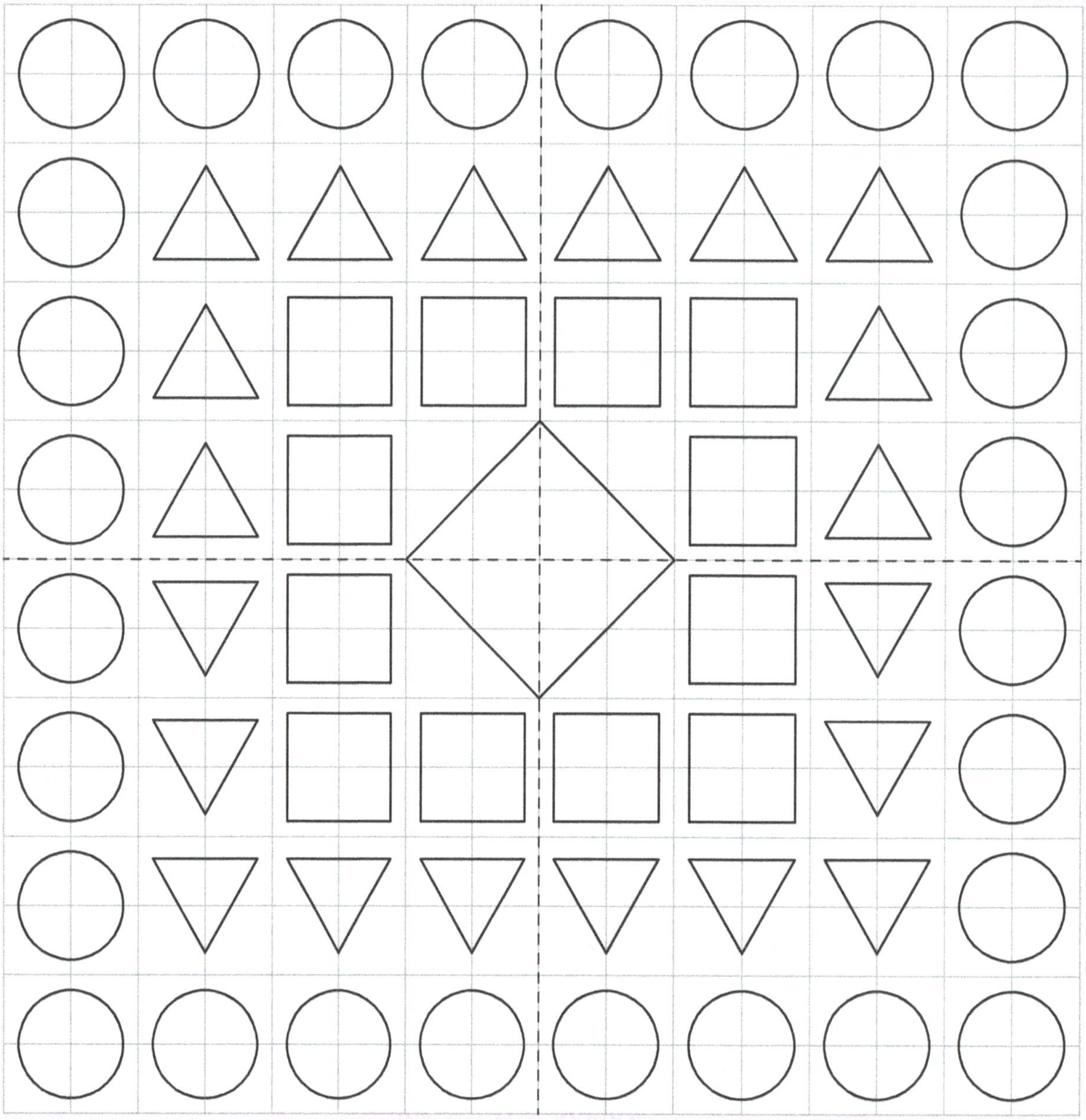

1 Write down what each cm represents using these scales.

a)

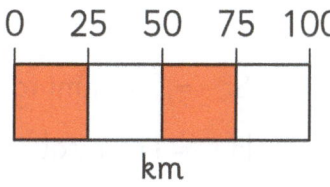

1 cm = ☐

b)

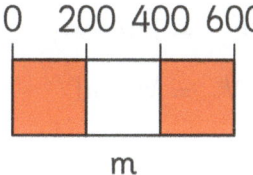

1 cm = ☐

c)

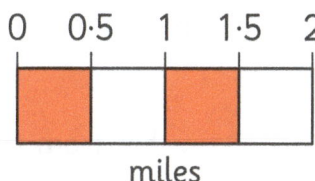

1 cm = ☐

d)

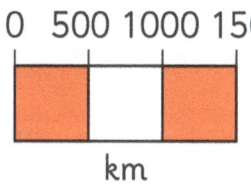

1 cm = ☐

2 Measure and complete these showing actual distance.

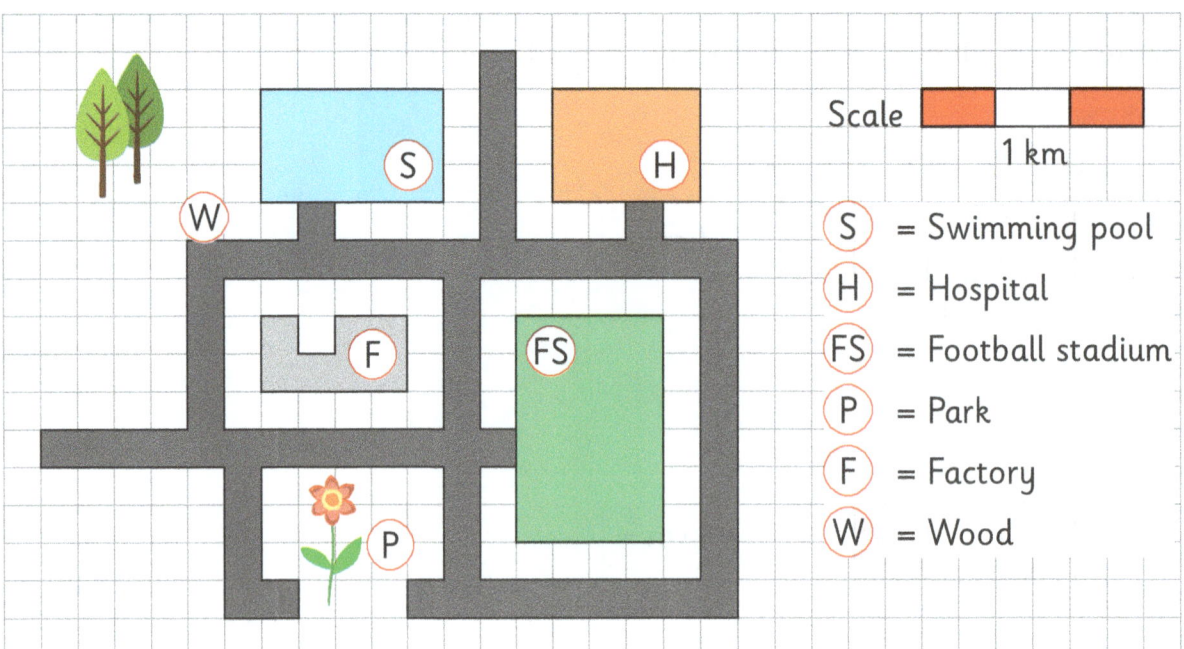

a) The entrance to the park is approximately [   ] km from the entrance to the woods.

b) The entrance to the hospital is approximately [   ] km from the entrance to the football stadium.

c) The entrance to the swimming pool is approximately [   ] km from the entrance to the woods.

d) The entrance to the swimming pool is approximately [   ] km from the entrance to the hospital.

e) The entrance to the hospital is approximately [   ] km from the entrance to the park.

f) The entrance to the [   ] is approximately 7 km from the entrance to the woods.

Use the space below to draw an animal park. Decide on your scale and mark it in.

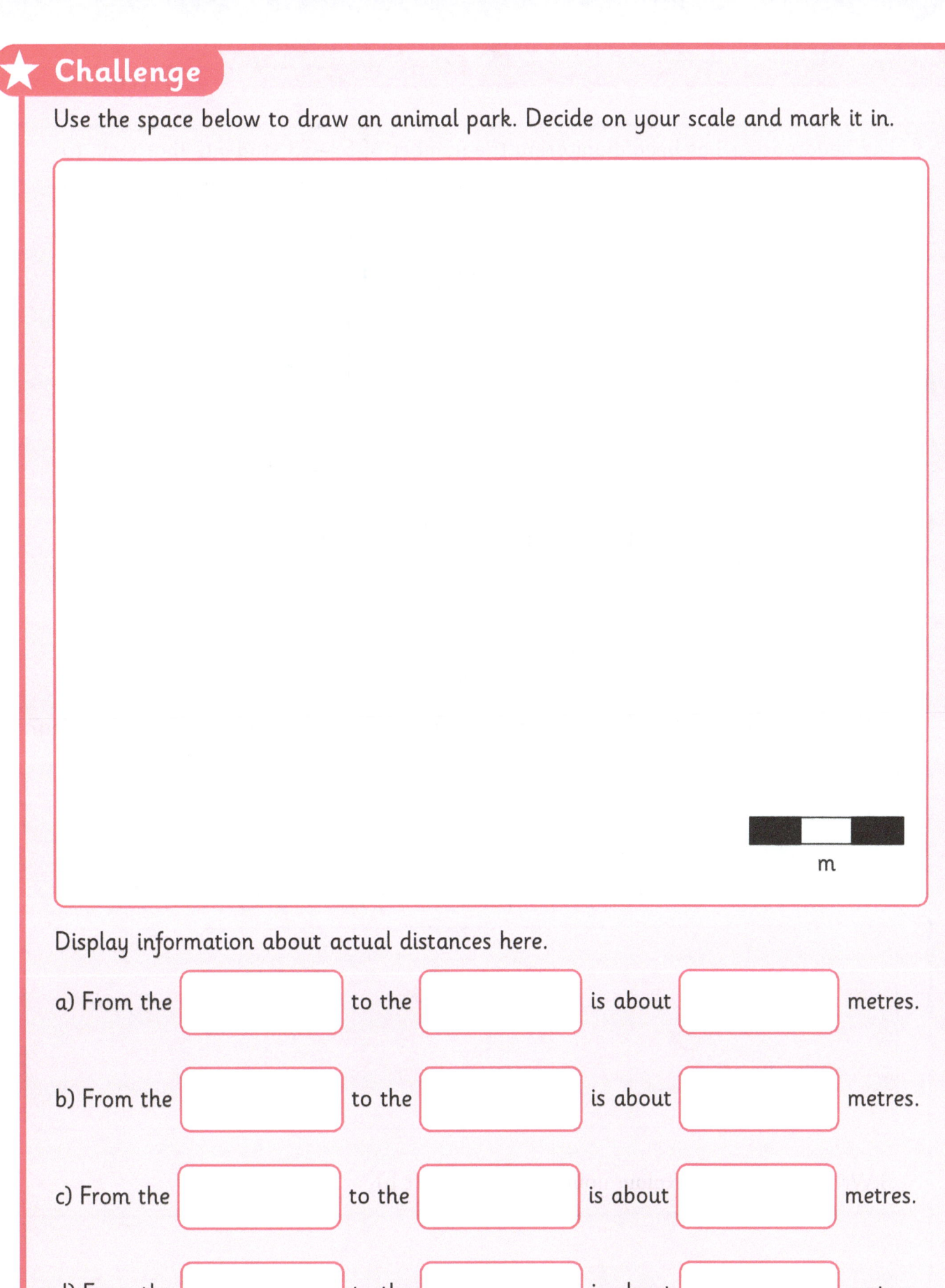

m

Display information about actual distances here.

a) From the [   ] to the [   ] is about [   ] metres.

b) From the [   ] to the [   ] is about [   ] metres.

c) From the [   ] to the [   ] is about [   ] metres.

d) From the [   ] to the [   ] is about [   ] metres.

e) From the [   ] to the [   ] is about [   ] metres.

1 The headteacher has been monitoring the number of S1–S3 students that walk to school in February and May. Her results are shown in the double bar graph below. Answer the questions that follow.

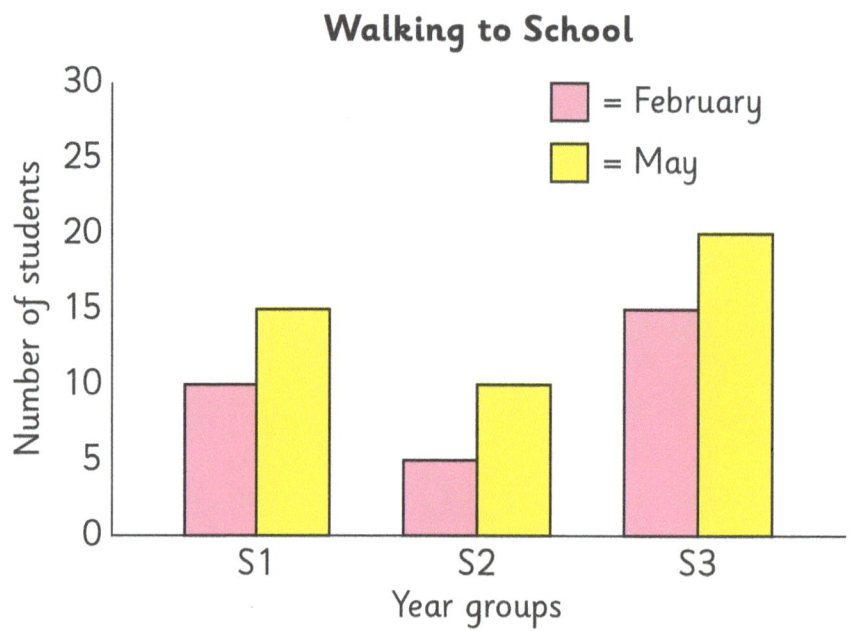

a) Which year group shows most students walking to school in both February and May?

b) What do you notice when you compare the data for February and May?

c) Write a reason to explain your answer to part b)

2 Cate and Katherine have been selling badges to raise money. This double line graph shows how each girl got on. Answer the questions below the graph.

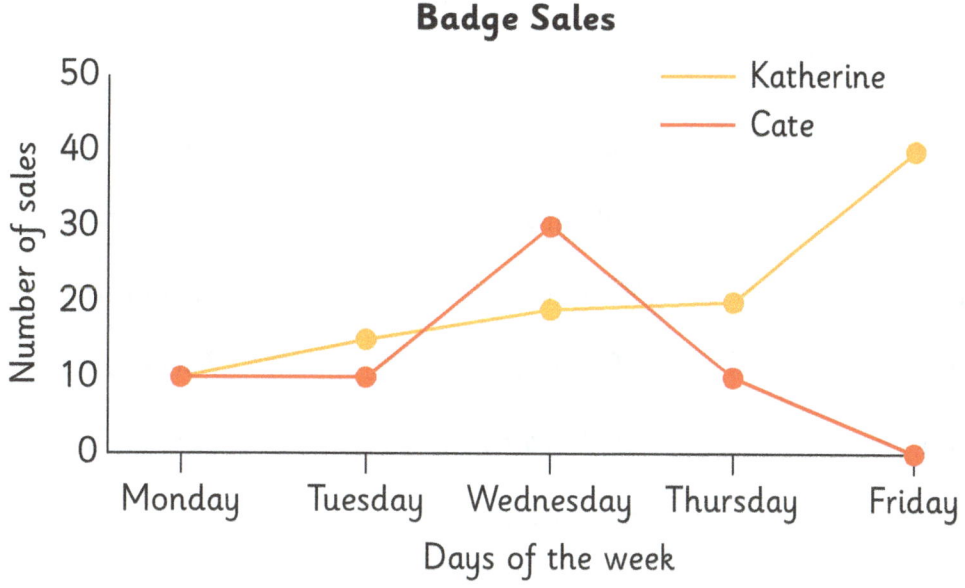

a) On Wednesday how many more badges did Cate sell than Katherine?

b) Who sold the most badges?

c) What reason might there be for Friday's sales?

S2 and S3 weighed the amount of food waste they had after lunch for a week. Their information is shown in the table. Look at this data and draw a double bar graph to display the information.

Day	S2	S3
Monday	950 grams	800 grams
Tuesday	900 grams	700 grams
Wednesday	900 grams	750 grams
Thursday	800 grams	350 grams
Friday	800 grams	300 grams

Write a question about this graph and ask a partner to answer it.

1 There are 24 students in Nuria's art class. They are buying lunch from the shop. 12 of them chose cheese sandwiches, 6 of them chose ham sandwiches, 3 of them chose tuna sandwiches and 3 of them chose egg sandwiches. Colour and label this pie chart to show the information.

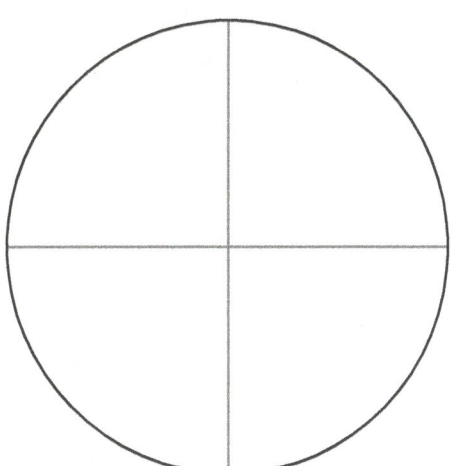

2 There are 36 students in the school talent show.

18 of them sing.

9 of them dance.

5 of them do comedy.

4 of them play guitar.

The pie chart has been divided into 36 sections. Each student is represented by

 °

Colour and label the pie chart to show the information.

a) What do half of the students do?

b) What do one quarter of the students do?

c) One  of the students play the guitar.

3 18 people were asked about the activities they did in their free time.

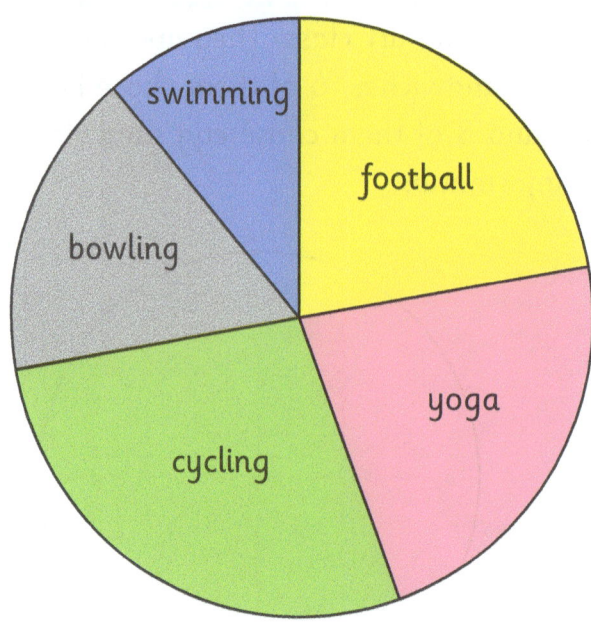

Measure each section to answer the questions. **Hint: 360 ÷18 = 20°**

a) Which is the most popular activity?

b) How many people go swimming?

c) How many more people go bowling than go swimming?

d) Which sports are equally popular?

Max's mum goes shopping for gifts for her family. Look at the pie chart and answer the questions.

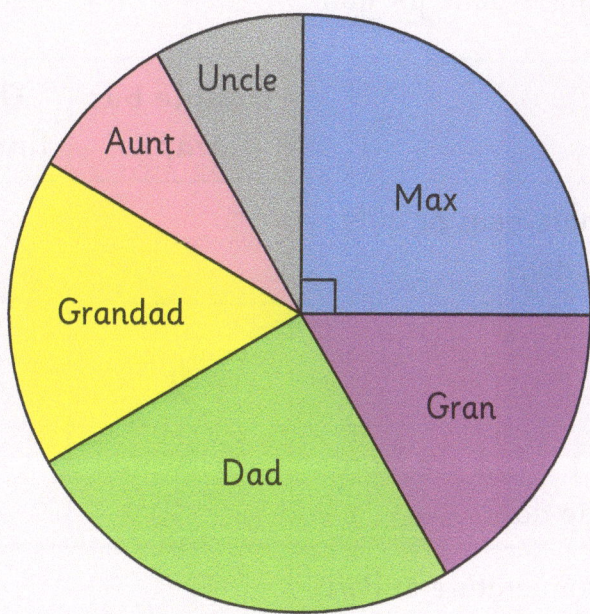

a) What fraction of the money did she spend on Max?

b) Max's mum spent **£40 on Gran**. Use this information to calculate how much she spent on everyone else.

Dad:

Grandad:

Aunt:

Uncle:

Max:

15.3 Creating and interpreting graphs

1 Decide which of these graphs would be best to display the data gathered from these questions. The first one is done for you.

Question	Double bar graph	Double line graph	Pie chart
I wonder whether more boys or girls go to First Aid each day?			Yes
I wonder what subjects my classmates like best?			
I wonder whether S1 or S2 use more water throughout the day?			
I wonder how the temperature in Paris compares to Glasgow?			
I wonder which of the five new films is most popular at the cinema today?			

2 Look at this pie chart. It shows the flowers used by Mrs Green, the florist, in April. Write true or false beside each statement.

April Flowers

a) "I notice that orchids are the most popular flower."

b) "I notice that orchids and lilies make up half of all the flowers used."

c) "I notice that lilies are the least popular flower."

d) "I notice that the number of lilies and daffodils are about the same as the number of tulips."

3 Look at this line graph. Write down three "I notice…" statements about it.

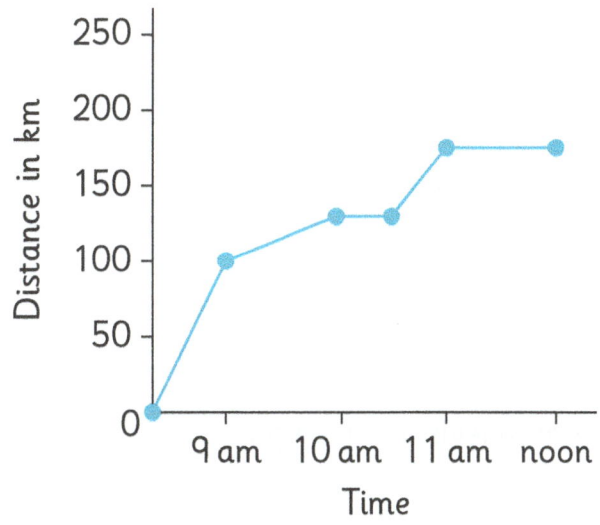

Journey of a delivery driver

"I notice [ ]"

"I notice [ ]"

"I notice [ ]"

Write down two "I wonder...." questions that you could ask each of these groups.

a) The people you live with.

I wonder

I wonder

b) The people you are in class with.

I wonder

I wonder

c) The people you see on television or on film.

I wonder

I wonder

1 a) Look at this bar graph. Decide if the statements are appropriate for the data
   shown. Circle yes or no.

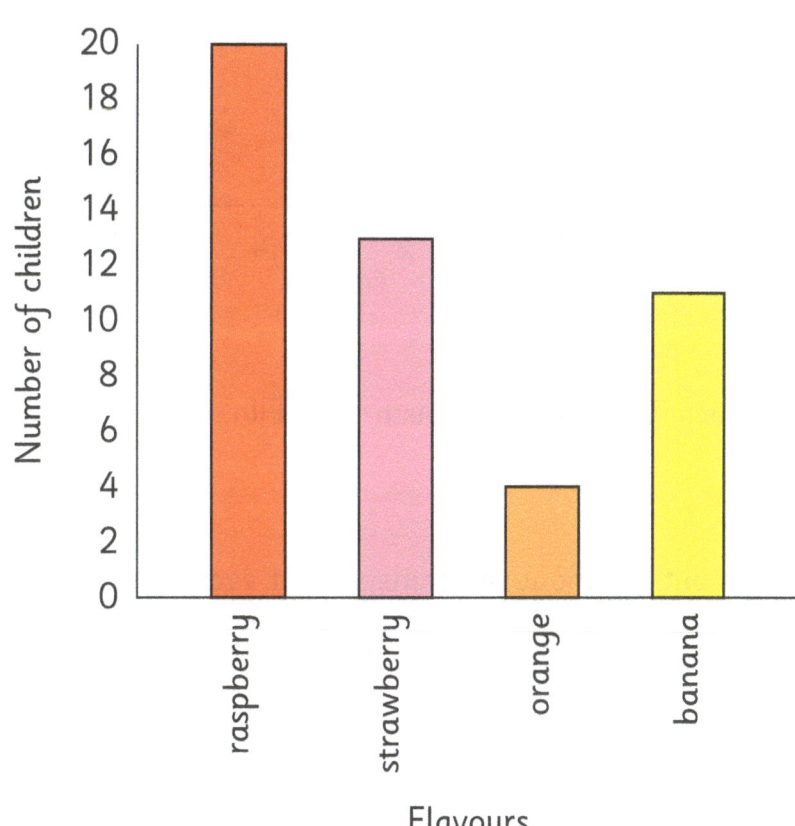

Favourite ice lolly flavours

i) I found that red fruits are most popular.   **YES**   **NO**

ii) I wonder why banana lollies are so unpopular.   **YES**   **NO**

iii) I found that people would have liked to have lime as a choice. **YES**   **NO**

iv) I think that the children who were asked were very young.   **YES**   **NO**

b) Now write an appropriate statement of your own about this graph.

2 a) Look at this line graph. Decide if the statements are appropriate for the data shown. Write yes or no.

Baby Lucy's weight from birth

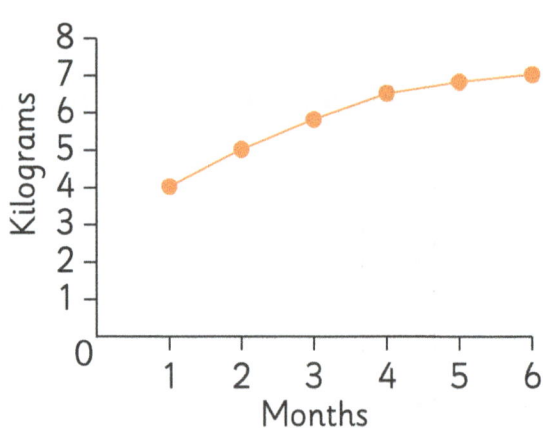

i) I wonder if Lucy was heavier at birth than Ella.

ii) I think babies put on about the same weight every month.

iii) I found that Ella almost doubled in weight in six months.

iv) I wonder why Lucy lost weight.

b) Write one appropriate statement and one inappropriate statement for this inquiry.

3 Look at this pie chart. It shows the patterns that people would like for their new sofa. Decide if the statements are appropriate for the data shown. Circle yes or no.

Sofa Choices

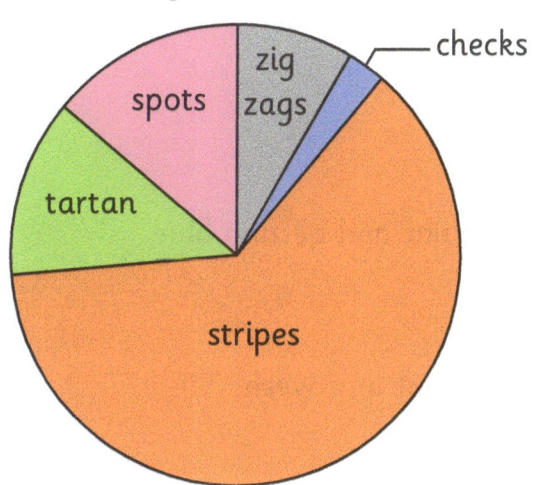

a) I found that more people would like stripes than any other pattern.   **YES**   **NO**

b) I wonder why checks and tartan have such similar results.   **YES**   **NO**

c) I found that less than 10 people like checks.   **YES**   **NO**

d) I think that the results would be different if people were allowed to choose no pattern.   **YES**   **NO**

⭐ **Challenge**

Think of five different questions that you are really interested in. You may not be able to carry out the investigation but your question should be able to be answered by data. An example is given:

• I wonder if more men or women go to my local gym before 9 am each day?

1 Write the words **certain**, **likely**, **even chance**, **unlikely** or **impossible** to describe the likelihood of these events occurring:

a) Tossing a coin and getting a tail

b) Mixing red and yellow paint and getting blue

c) Going to sleep at some point in a week

d) Meeting a prince

e) Growing a banana tree from an apple pip

f) Seeing a red car on the journey home from school

g) Sitting with a friend at lunch time

2 Ceren has a bag with 10 shapes in it.

Pick from the words **certain**, **likely**, **even chance**, **unlikely** or **impossible** to say how likely each of these is:

a) Ceren pulls out a pentagon.

b) Ceren pulls out a rectangle.

c) Ceren pulls out a shape with more than three sides.

3 Alex has a six-sided spinner. Write the percentage chance – 0%, 25%, 50%, 75% and 100% – to say how likely these are:

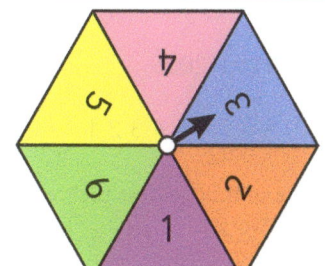

a) Alex spins an even number.

b) Alex spins a number less than 9.

4 Write 0, $\frac{1}{4}$, $\frac{1}{2}$, $\frac{3}{4}$, or 1 to show how likely these are:

a) Dougie picks an apple.

b) Dougie picks a yellow fruit.

★ **Challenge**

Colour the cubes so that the following are true.
Your choices are red, purple, green, yellow, orange and blue.

- I have a 100% chance of drawing a colour with an **e** in the name.
- I have an even chance of drawing a colour with an **l** in the name.
- I have a 0·25 chance of drawing a colour with three vowels in the name.
- I have 0 chance of drawing a colour with a **d** in the name.

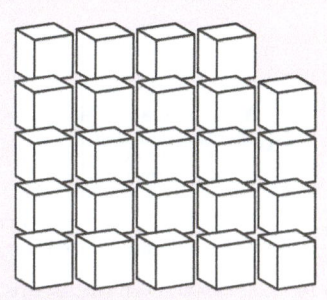

© 2025 Leckie

001/16012025

10 9 8 7 6 5 4 3 2 1

ISBN 9780008741341

Published by
Leckie
An imprint of HarperCollins Publishers
Westerhill Road, Bishopbriggs, Glasgow, G64 2QT

T: 0844 576 8126 F: 0844 576 8131
leckiescotland@harpercollins.co.uk www.leckiescotland.co.uk

HarperCollins Publishers
Macken House, 39/40 Mayor Street Upper, Dublin 1, D01 C9W8, Ireland

Publisher: Fiona McGlade

Special thanks
Project editor: Peter Dennis
Layout: Jouve
Proofreader: Louise Robb

A CIP Catalogue record for this book is available from the British Library.

Acknowledgements
Images © Shutterstock.com

Printed in United Kingdom.